Global Web3
Eco Innovation

Singapore University of Social Sciences - World Scientific Future Economy Series

ISSN: 2661-3905

Series Editor
David Lee Kuo Chuen *(Singapore University of Social Sciences, Singapore)*

Subject Editors
Guan Chong *(Singapore University of Social Sciences, Singapore)*
Ding Ding *(Singapore University of Social Sciences, Singapore)*

Singapore University of Social Sciences - World Scientific Future Economy Series introduces the new technology trends and challenges that businesses today face, financial management in the digital economy, blockchain technology, smart contract and cryptography. The authors describe current issues that the business leaders and finance professionals are facing, as well as developments in digitalisation. The series covers several increasingly important new areas such as the fourth industrial revolution, Internet of Things (IoT), blockchain technology, artificial intelligence (AI) and many other forces of disruption and breakthroughs that shape today's realities of the economy. A better understanding of the changing environment in the future economy can enable business professionals and leaders to recognise realities, embrace changes, and create new opportunities — locally and globally — in this inevitable digital age.

Published*

Vol. 8   *Global Web3 Eco Innovation*
          by DeFiDAO, David Lee Kuo Chuen, Guan Chong and Ding Ding

Vol. 7   *Inclusive Disruption: Digital Capitalism, Deep Technology and Trade Disputes*
          by David Lee Kuo Chuen, Linda Low, Joseph Lim and Carmen Shih Chia Mei

Vol. 6   *Financial Management in the Digital Economy*
          edited by David Lee Kuo Chuen, Ding Ding and Guan Chong

Vol. 5   *The Digital Transformation of Property in Greater China: Finance, 5G, AI, and Blockchain*
          by Paul Schulte, Dean Sun and Roman Shemakov

Vol. 4   *Blockchain and Smart Contracts: Design Thinking and Programming for FinTech*
          by Lo Swee Won, Wang Yu and David Lee Kuo Chuen

**More information on this series can also be found at
https://www.worldscientific.com/series/susswsfes*

(Continued at end of book)

Singapore University of Social Sciences - World Scientific
Future Economy Series : **8**

Global Web3 Eco Innovation

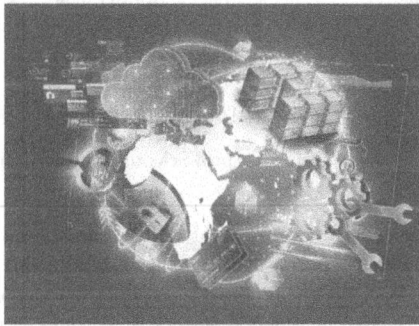

DeFiDAO

David LEE Kuo Chuen
Singapore University of Social Sciences, Singapore

GUAN Chong
Singapore University of Social Sciences, Singapore

DING Ding
Singapore University of Social Sciences, Singapore

SUSS
SINGAPORE UNIVERSITY
OF SOCIAL SCIENCES

World Scientific

Published by

World Scientific Publishing Co. Pte. Ltd.

5 Toh Tuck Link, Singapore 596224

USA office: 27 Warren Street, Suite 401-402, Hackensack, NJ 07601

UK office: 57 Shelton Street, Covent Garden, London WC2H 9HE

Library of Congress Control Number: 2023043113

British Library Cataloguing-in-Publication Data
A catalogue record for this book is available from the British Library.

Singapore University of Social Sciences - World Scientific Future Economy Series — Vol. 8
GLOBAL WEB3 ECO INNOVATION

Copyright © 2024 by World Scientific Publishing Co. Pte. Ltd.

ISBN 978-981-128-367-3 (hardcover)
ISBN 978-981-128-452-6 (paperback)
ISBN 978-981-128-368-0 (ebook for institutions)
ISBN 978-981-128-369-7 (ebook for individuals)

For any available supplementary material, please visit
https://www.worldscientific.com/worldscibooks/10.1142/13605#t=suppl

Desk Editors: Soundararajan Raghuraman/Yulin Jiang

Typeset by Stallion Press
Email: enquiries@stallionpress.com

Preface

Along with the maturing of blockchain technology, we have seen the border of Web3 expanding from hash rate to crypto market, then from crypto market to metaverse. Our book starts by introducing the origin of the Web3 concept; Then, we look into the infrastructure of Web3, namely the blockchain and its main applications — the development of which started from the genesis block of BTC to date. The book also covers the key developing tracks of the current Web3 world, including DeFi, NFT, GameFi and DAO. After that, it is the forward-looking chapter on Metaverse, which includes some analysis that could only be based on reasonable estimation. The last chapter is a review of the "twins" of Web3 — investors and regulators.

About the Editors

DeFiDAO Founded in 2020, DeFiDAO has grown to become one of the most professional and insightful self-publishing houses in the crypto space in China. DeFiDAO is dedicated to exploring web 3.0's cutting-edge gameplay and innovative design to provide forward-looking content for the crypto community.

David Lee Kuo Chuen is a Professor of Fintech and Blockchain at the Singapore University of Social Sciences (SUSS), a Distinguished Professor at Shanghai University of Finance and Economics, an Adjunct Professor at the National University of Singapore (NUS), a founding Council Member of the British Blockchain Association, and Editor-in-Chief or Deputy Editor-in-Chief of the *Journal of Fintech* and *Annual Review of Fintech*.

Meanwhile, Professor Lee is also an internationally renowned scholar and industry opinion leader in the field of blockchain, an expert in the development of Web3 and Artificial Intelligence (AI) application scenarios, the Chairman of the Global FinTech Institute (GFI), a co-founder of the Singapore Blockchain Association, a co-founder of the Blockchain Security Alliance and Global Web3 Association, a CBDC advisor to the Asian Development Bank, Vice President of the Economic Society of Singapore (ESS), a cryptocurrency advisor to the Asian Institute of Digital Finance (AIDF) of NUS, an advisor to the SUSS Node for inclusive FinTech (NiFT), an independent director of several listed companies in

Singapore, an angel investor in blockchain, Web3, inclusive finance and AI innovation, a senior advisor and Investment Committee member of Artichoke Capital which is backed by institutional limited partners, including sovereign wealth and family office funds.

Professor Lee previously served as a Senior Advisor to the Provost of Singapore Management University (SMU), Director of the Sim Kee Boon Institute for Financial Economic (SKBI) in Singapore, a member of the Research Council of the Monetary Authority of Singapore (MAS), a member of the Technical Advisory Committee of Ping An OneConnect, an independent director of Lu Global in Singapore, and long taught at the Mayor's Class in Singapore. In addition, Professor Lee served as an international consultant to the Distributed Trust Research Project at Stanford University, Group MD of listed companies OUE Enterprise and Auric Pacific, a Fellow at Harvard University – National University of Singapore (1991–1993), a Fulbright Scholar at Stanford University in 2015, the founder of hedge fund Ferrell Asset Management, and the founder of real estate developer Ferrell Residences.

Chong Guan is Associate Professor and Director in Centre for Continuing and Professional Education, Singapore University of Social Sciences (SUSS), Singapore. Her research interests lie in the area of data-driven digital marketing. Her publications appear in leading journals such as *Electronic Markets, European Journal of Marketing, Journal of Interactive Marketing, Journal of Business Research and Telecommunications Policy.* She is an editorial board member of *Internet Research.* She has consulted business practices and conducted corporate training sessions on social media analytics, AI in marketing, and other emerging areas.

Ding Ding is the Vice Dean of the School Business at Singapore University of Social Sciences (SUSS) where she has worked since 2008. She received her PhD in Economics from Nanyang Technological University and is a Chartered Financial Analyst (CFA). Prior to her current role, she was the head of BSc in Finance programme at SUSS during 2012–2018 and taught various finance courses at the undergraduate, post-graduate, and executive training level. Her recent research focuses on the developments and applications of financial technology (FinTech), and she has published several books and articles in this area.

About the Contributors

Bosheng Ding is a researcher and PhD candidate at Alibaba DAMO Academy and Nanyang Technological University, with a focus on deep learning and natural language processing. He completed his undergraduate degree in Electrical and Electronic Engineering at Nanyang Technological University in Singapore. Bosheng has a deep understanding of Generative AI, including Stable Diffusion and ChatGPT. He has published multiple papers at top-tier artificial intelligence conferences and has served as a reviewer for various AI conferences, including EMNLP, ACL, and WSDM. In addition, he is a visiting lecturer on digital marketing at the Singapore University of Social Sciences. Before pursuing his PhD, Bosheng worked as a consultant at PwC Singapore, where he served clients in various large financial institutions across countries such as Singapore, Thailand, the Philippines, Malaysia, South Africa, and Australia.

Brian Heng is Associate Professor in the School of Business and former Director of Centre for Professional and Continuing Education (CCPE) in Singapore University of Social Sciences. He led CCPE to develop and execute continuing education and training (CET) strategies in delivering industry relevant programmes, courses, and training on social sciences, business management, and technology. With more than 20 years of experience in marketing, business development, corporate strategy, and development, as well as human resources, Brian has also been deeply involved in greenfield projects across retail, insurance, education, oil and gas industries. His research interests include the digital economy, marketing and communications, entrepreneurship, overseas Chinese capitalism, future skills, future work, and the knowledge Brian has designed and curated many courses, events, and workshops that aims to support the needs of individual professional development, organizational learning and development, and national preparedness for the future economy.

Gavin Qu is a partner of 8BTC and the CEO of DeFiDAO. He is a Web enthusiast, an early observer of the crypto market, and a veteran in the blockchain sector, having led a number of innovative blockchain and community projects. He also co-authored the *2014–2015 Cryptocurrency Industry Report*, and his research interest spans all major Web3 areas including DeFi, NFT, Game, DAO, and metaverse.

Jiajian Wang is a veteran journalist. In the past four years, he has interviewed more than 100 blockchain practitioners, written more than 500 articles including event reports, trend analysis, and interviews. He is the author of *2021 Q1 Blockchain Supercomputing Industry Report*, exploring the development status and future trends of the computing industry.

Jiancang Guo obtained his PhD from the Department of Mathematics at University of Singapore in 2017. He is currently a senior research scientist at Netvirta Singapore. His research interest is in applied mathematics and computer vision, including the analysis and simulation of rare events, image recognition, and 3D modeling.

Jincheng Zheng (Jesse) is a research fellow at the Singapore University of Social Sciences (SUSS). He is a Web3 enthusiast who deeply dives into DeFi, NFT, and DAO. Before joining SUSS, he was a research associate at the economics department of Nanyang Technological University (NTU) and a key opinion leader from a leading crypto community in Asia. Jesse Zheng is a graduate of the Master of Science (Managerial Economics) programme at NTU.

Kyle Zhao, a crypto enthusiast, has been all in crypto since 2016, loving to share with the crypto community the latest news on DeFi, NFT, gaming, Web3, and so forth. Recently he has focused on scaling solutions (e.g., Ethereum rollups, Celestia modular solution) and Web3.

Mozhi Wang completed MSc in Finance of John Molson Business School, Concordia, Canada. As a researcher and analyst at Timestamp Capital, the investment branch of DeFiDAO. He built the new valuation model for Timestamp Capital and designed a series of DeFi protocols on Solana, and his research efforts cover all important areas of crypto applications (DeFi, NFT, Game, DAO, infra, and metaverse), with continuous tracking of crypto markets and global regulations.

Qinxu Ding is a lecturer at the Singapore University of Social Sciences. Prior to joining SUSS, he was a research fellow at Alibaba–NTU Singapore Joint Research Institute, writing academic AI papers, and applying the research outcomes for Alibaba industry applications. Dr Ding completed his PhD in EEE at Nanyang Technological University. His research interests include Explainable AI, Robust AI, Recommendation Systems, Blockchain, and Computational Mathematics. He has been teaching undergraduate level Fintech courses. His publications appear in Tier-1 AI conferences such as NeurIPS, AAAI, and CIKM. He has also published a series of computational mathematical journals such as *Mathematics Methods in the Applied Sciences, Numerical Methods for Partial Differential Equations, International Journal of Numerical Methods in Fluids, Advances in Difference Equations, Boundary Value Problems,* and conferences such as ICARCV and ICNAAM. He is also a core team member of SUSS Node for Inclusive Fintech (NiFT).

Richard Li is the Deputy Managing Editor at DeFiDAO. He worked for many years as the editor managing a Hong Kong-based international bilingual journal that focuses on legal and business analysis for corporate leaders and general counsel.

Sue Tang is an experienced Web3 builder, a DAO enthusiast, and a writer for the blockchain media DeFiDAO. She has been active in several DAOs and, in particular, she is the core contributor to DAOrayaki and SeeDAO. With a master's degree in British and American literature and three years of experience in the blockchain industry, Sue is also a veteran translator in Web3, committed to bridging the gap between Chinese and foreign cultures in this space.

Swee Won Lo is a senior lecturer at the Singapore University of Social Sciences (SUSS) and the Deputy Director of the SUSS Node for Inclusive FinTech (NiFT). Her teaching expertise is in blockchain technologies and applications, smart contracts, financial cryptography, and blockchain/data security and privacy. Her research interests include applied cryptography, blockchain scalability, risks, and sustainability. She is a member of the Blockchain and Distributed Ledger Technologies Technical Committee at ITSC, a Non-Executive Director at Coin Ultimate Trading, and she serves as a review editor for the Frontiers in Big Data (Cybersecurity and Privacy), associate editor for the *Journal of Digital Assets*, and is a member of the Centre for Evidence-Based Blockchain by the *Journal of British Blockchain Association*.

Wendy Wang is the editor of DeFiDAO. Having entered the crypto world since 2016, she remains passionate about exploring new things, and she always love learning Web3.

Xueyuan Qin is the editor of DeFiDAO and has been working for the blockchain media since 2017. He has been writing industry-related articles, with research interests in emerging areas such as DeFi, DAO, and NFT.

Yu-Chao Cheng is a senior lecturer at the Singapore University of Social Sciences (SUSS) School of Business. He was a member of the expert group for Singapore IP Strategy 2030 (SIPS2030), which oversaw "Driving Enterprises Growth with Intangible Assets". He specializes in Intellectual Property (IP) Policy, patent analysis management, competitive intelligence, and innovation management. He was the investigator for various national projects on patent trends and litigation in the information and optical-electronic industries and carried out several consulting projects for businesses in various industries, including the semiconductor and financial leasing industries, IP deployment, and patent analysis.

Yu-chen Hung is a senior lecturer at the Singapore University of Social Sciences. Her teaching expertise is in digital platform strategy, integrated marketing communication, business-to-business marketing, consumer behavior, and global marketing. Her research interests include innovation adoption and diffusion, consumer psychology, and experiential marketing. Her works have appeared in various international journals across disciplines, such as *International Journal of Research in Marketing, European Journal of Marketing, Journal of Business Research, Electronic Markets, and Transportation,* and *Journal of Marketing Management.*

Contents

Chapter 1

Web3: The Renaissance of Cyberspace

Gavin Qu and Bosheng Ding

Beginning in late 2021, the search volume of the keyword "Web3" on the Internet has grown rapidly. People started talking about Web3, and it seemed that the Web3 dream could become a reality tomorrow.

Web3 is not a product that appeared out of thin air but a continuation of the spirit of cyberpunk and cypherpunk in the 1980s and 1990s. The current Web3 revolution resembles a renaissance with the economic stimulus flowing into cyberspace.

1.   The Declaration of Independence of Cyberspace

On February 8, 1996, the founder of the Electronic Frontier Foundation, John Perry Barlow, issued *A Declaration of Independence of Cyberspace*, declaring that Cyberspace is a new home of Mind and a global social space naturally independent of the tyrannies imposed by governments of the industrial world (Barlow, 2019).

The declaration mainly embodies the following three propositions:

1. **Matterless**: "Ours is a world that is both everywhere and nowhere, but it is not where bodies live".
2. **Borderless**: "We are creating a world that all may enter without privilege or prejudice accorded by race, economic power, military force, or station of birth".

3. **Non-discriminatory**: "We are creating a world where anyone, anywhere may express his or her beliefs, no matter how singular, without fear of being coerced into silence or conformity".

Barlow's manifesto quickly became well-known and widespread on the Internet. Nine months after publication, it received about 40,000 website retweets.

> *We will create a civilization of the Mind in Cyberspace.*
>
> — John Perry Barlow

However, with the development of the Internet, his manifesto has been questioned by more and more people, including himself. In 2002, the number of sites reproducing the manifesto had dropped to approximately 20,000. In a 2004 interview, Barlow reflected on the work he did in the 1990s, particularly the optimism he held at the time. "We all get older and smarter", he said. Obviously, the scene depicted in the declaration was not realized at the time, but this did not affect the continuous pursuit of Cyberspace by its believers.

2. Early Attempts at Cyberspace Sovereign Currency

If a currency is a stimulus for the efficient operation of modern economic society, cyberspace independent of the physical world should also have a native currency system and carry out economic activities accordingly.

The Cypherpunk movement was thriving almost concurrently with the birth of *A Declaration of Independence of Cyberspace*. In *A Cypherpunk's Manifesto* published by Eric Hughes in 1993, the vision and mission of the cypherpunks were stated, which is to build an anonymous system through methods such as cryptography to defend people's privacy (Hughes, 1993). At the same time, the manifesto also mentioned that "software can't be destroyed and that a widely dispersed system can't be shut down".

We the Cypherpunks are dedicated to building anonymous systems. We are defending our privacy with cryptography, with anonymous mail forwarding systems, with digital signatures, and with electronic money.

— Eric Hughes

In 1983, David Chaum proposed an anonymous electronic cash system based on the blind signature technology, which is the predecessor of the electronic currency eCash. However, the system didn't survive, as the operating company behind it, DigiCash, declared bankruptcy in 1998. One could explain DigiCash's failure with many possible reasons, but one of the primary reasons is its centralized architecture: once the centralized company or server crashes, the system will be unsustainable. It is hard to imagine that, in the future, we will use a company's product as a common currency for the Internet to conduct transactions.

In the same year that DigiCash collapsed, another cypherpunk, Wei Dai, proposed b-money, an anonymous and distributed electronic cash system. B-money has the basic characteristics of all modern cryptographic currency systems. However, due to various technical implementation reasons, b-money has never been officially launched.

In 2005, Nick Szabo designed a decentralized digital currency mechanism called bit gold. All data in cyberspace can be easily copied and pasted, which means that the design of digital currency needs to solve the "double spending problem". Most digital currencies solve this problem by introducing a centralized authority to record the balances of all accounts, but Szabo does not agree with this solution: "I was trying to mimic as closely as possible, in cyberspace, the security and trust characteristics of gold, and chief among those is that it doesn't depend on a trusted central authority". From eCash to b-money to bit gold, there were many attempts made by the early cypherpunks to create the native sovereign currency of the cyberspace, but none of them were eventually adopted.

3. Software is Eating the World

At the same time, the Internet has also completed the transition from the Web 1.0 era to the Web 2.0 era, but it has also encountered development bottlenecks that are difficult to solve with the existing architecture.

Web 1.0 is a retrospective term that refers to the first phase of the development of the World Wide Web, which ran from approximately 1991 to 2004. At this stage, there were very few content creators, and the vast majority of users were simply consumers of content.

In the era of Web 2.0, general Internet users can exchange information and collaborate on various Internet platforms at a very low cost. At this time, the core concepts of Internet products were interaction, sharing, and association. During this period, in 2011, a16z partner Marc Andreessen devised the famous slogan: "software is eating the world" (Andreessen, 2011). "We believe that many of the prominent new Internet companies are building real, high-growth, high-margin, highly defensible businesses," he wrote.

After that, we did see the rapid rise of tech giants such as Meta (formerly Facebook), Amazon, Alphabet (Google's parent company), and Tencent. Although the core functions of these giants are different, one thing in common in their rise is the ability to obtain "state" from users.

In computer systems, "state" refers to the state of a thing at a certain time, while stateful means that, given the same input, the output value will change according to the state at different points in time. For example, a user uses the search service provided by Google, and each click on the search results page can help the search engine to provide more accurate search results for the next user.

In the era of Web 2.0, users are not only users of Internet services but also part of Internet products. The "state" of Internet services grows compounded, and users trust the platform and give their "states" in exchange for better services. At the same time, platform service providers also have higher valuations.

But at the end of the honeymoon period, as the growth of the platforms enters a bottleneck period, they will often fail the trust of users, and the relationship with users will change from a positive-sum relationship to a zero-sum relationship. Platforms need to extract all kinds of data, including private data, from users to keep growing, turning their former partners into competitors. At the same time, the Internet platform has obtained a very high "state" barrier through the accumulation of "states" over the years, which is unclimbable for new entrepreneurs, hindering the emergence of competition and innovation.

Software is eating the world, and the services above the software are beginning to erode the interests of participants. Thus, the Internet needs a paradigm shift urgently.

4. Blockchain Genesis

On October 31, 2008, Satoshi Nakamoto published the Bitcoin white paper on a cypherpunks mailing list (see Figure 1). And two months

Bitcoin P2P e-cash paper

Satoshi Nakamoto satoshi at vistomail.com
Fri Oct 31 14:10:00 EDT 2008

- Previous message: Fw: SHA-3 lounge
- **Messages sorted by:** [date] [thread] [subject] [author]

```
I've been working on a new electronic cash system that's fully
peer-to-peer, with no trusted third party.

The paper is available at:
http://www.bitcoin.org/bitcoin.pdf

The main properties:
Double-spending is prevented with a peer-to-peer network.
No mint or other trusted parties.
Participants can be anonymous.
New coins are made from Hashcash style proof-of-work.
The proof-of-work for new coin generation also powers the
    network to prevent double-spending.

Bitcoin: A Peer-to-Peer Electronic Cash System
```

Figure 1.   Bitcoin P2P e-cash paper.

later, on January 3, 2009, the genesis block of Bitcoin was mined. This marks the advent of the trust-free Internet-native currency that the cypherpunks have been seeking for the past few decades, and cyberspace has gained a fresh stimulus of economic activity.

On January 24, 2014, Vitalik Buterin officially announced the birth of the Ethereum project at the Miami Bitcoin Conference. Compared with Bitcoin, Ethereum provides developers with higher flexibility: Ethereum introduces a Turing-complete virtual machine into the blockchain, turning the entire network into a universal virtual computer shared by the world. The emergence of DeFi protocols such as Uniswap and Compound means that people can engage in more and more complex business activities, such as transactions and lending, in cyberspace. After this, the emergence of new assets, such as NFT, GameFi, and DAO, will also provide more venues for the natives in cyberspace.

In April 2014, the co-founder and former CTO of Ethereum, Gavin Wood, first systematically expounded the concept of Web3. Gavin believes that in the post-Snowden era, Internet users can no longer continue to trust companies, and companies will only manage and use user data for their own profit-making purposes. Therefore, it is necessary to build trust-minimized Internet infrastructures and applications. According to Gavin, "Web 3.0 is an eclectic set of protocols that provide application developers with the building blocks to build applications in entirely new ways. These technologies enable users to verify the (authenticity) of information received and sent, ensuring that they are processed in a transaction. Reliably give and receive in the process. Web 3.0 can be seen as an enforceable Magna Carta of the Internet and a cornerstone of individual liberty against authority".

So far, the resurgent cyberspace has taken shape, which will be a decentralized network system:

1. Open and verifiable where all participants control the system and own their states;
2. Inclusive and non-discriminatory, where all participants have equal rights to use the network;

3. There is no single point of failure and the network structure is highly robust;
4. There is no centralized decision-making governance and all changes require the authorization of the majority of participants;
5. Cyberspace has a native trust-free economic system.

The thriving community of decentralized autonomous organizations (DAOs) and Web3 applications has shown us the power of cyberspace where strangers on the Internet come together based on the same values and mission. Moreover, as the infrastructure evolves, countless possibilities remain to be discovered in the future.

Finally, I would like to conclude this chapter with a quote from Kyle Samani, co-founder of Multicoin Capital (Samani, 2019):

Trust is the foundation of all economic relationships. The greatest investment opportunity of our lifetimes is betting that it doesn't have to be.

— Kyle Samani

References

Andreessen, M. (2011). Why software is eating the world, *Wall Street Journal*, 20, C2 (2011).

Barlow, J. P. (2019). A declaration of the independence of cyberspace, *Duke Law & Technology Review*, 18(1), 5–7.

Hughes, E. (1993). A cypherpunk's manifesto. Accessed August 3 2004. http://www.activism.net/cypherpunk/manifesto.html.

Samani, K. (2019). Crypto Mega Theses.

Chapter 2

Infrastructure — Public Blockchains

Mozhi Wang, Xueyan Qin, Yu-chen Hung,
and Swee Won Lo

The Web3 revolution may have started long ago, but the epoch of blockchain history marked its beginning with the birth of Bitcoin in 2009.[1] The public chain is arguably the most important infrastructure in the short history of blockchain. The evolution of public chains has undergone three significant iterations in the past 13 years — from Bitcoin's Proof of Work (PoW) to Ethereum 1.0 with smart contract functionality and the various Layer 1 networks based on Proof of Stake (PoS). Today's Web3 is a hybrid system where the three models coexist and thrive in the ecosystem.

1. Bitcoin's Mystique

Bitcoin is entering the cycle of its fourth halving (Lavie, 2022; Waters & Maddrey, 2022). While the Bitcoin blockchain continues to grow, it is getting harder to pinpoint what a Bitcoin is. Too many layers of meaning have been given to the term "coin" since its genesis.

---

[1]The Bitcoin paper by Satoshi Nakamoto Nakamoto, S. (2008). *Bitcoin: A Peer-to-Peer Electronic Cash System.* https://bitcoin.org/bitcoin.pdf was published in 2008.

To disentangle the mystique around Bitcoin, this section inspects Bitcoin from various angles.

1.1. *Bitcoin vs fiat*

Bitcoin fanatics are still convinced that the cryptocurrency will replace fiat money as the world's universal payment instrument. According to the Bitcoin Whitepaper (Nakamoto, 2008), Bitcoin is a peer-to-peer payment system. Bitcoin fanatics' confidence in their belief has been strengthened by the fact that El Salvador first recognized BTC as one of the country's legal tender in September 2021 (Renteria *et al.*, 2021).

But this top-down rollout of Bitcoin was met with resistance from the public. Anti-Bitcoin protesters marched the streets; a significant portion of people downloaded the wallet for an initial $30 payout but stopped using it shortly after, as a few retail merchants accepted this payment form (McDonald, 2022).

El Salvador announced the issue of the first Bitcoin-backed bond worth $1 billion, or the "Volcano Bond," in March 2022 (Hetzber, 2022). The announcement is yet to be materialized to date. Other countries are also considering accepting bitcoin as legal tender, but only the Central African Republic made an official announcement (Kedem, 2022).

Could Bitcoin eventually replace fiat currency one day? Will Bitcoin ever replace the US dollar as the world's dominant currency? The Bank for International Settlements refuted such a possibility in the special edition of its annual economic report, "The Future Monetary System", released on June 12, 2022 (The Bank for International Settlements, 2022). Their view seems to be shared among most governments and monetary regulators worldwide.

However, the Bitcoin payment system and wallets will bring financial services to people who lack access to bank accounts. El Salvador might not have succeeded in making Bitcoin a nationally adopted currency. Yet, they have popularized the Bitcoin Lightning Network wallets, allowing many locals to receive USD remittances from their families overseas (Jenkinson, 2022). Thus the locals can avail one more choice of payment.

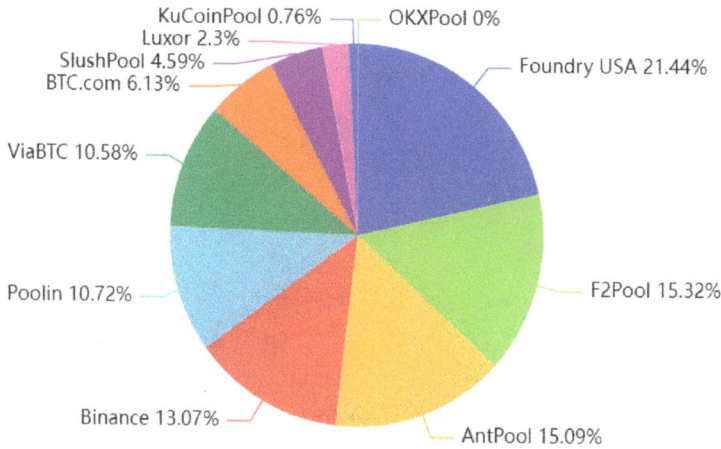

Figure 1. Global hashrate distribution.
Source: BTC123.fans.[2]

1.2. *Bitcoin vs assets*

Bitcoin has always been a "mine". But the days of individual "miners" are over. Instead, institutions have become the primary "miners" (see Figure 1 for the distribution of the miners).

Bitcoin's verification mechanism, Proof of Work (PoW), consumes energy to mine blocks by "solving cryptographic puzzles". Ethereum has adopted Proof of Stake (PoS) as its verification mechanism (this has been in Ethereum's original roadmap since its inception). The high energy consumption issue has led some countries to take a stern stand on Bitcoin mining in 2021. Some countries banned cryptocurrency trading, and the market turned bearish. Despite the change in trend, PoW has continued the momentum that lasts more than a decade.

Bitcoin has been cannibalizing the gold market for the past ten years (see Figure 2). Irrespective of market conditions, Bitcoin has become an asset class to hedge against risk. For instance, Ray Dalio and many investors have added a small percentage of Bitcoin to their

---

[2]BTC123.fans. https://history.btc123.fans/hashrate/.

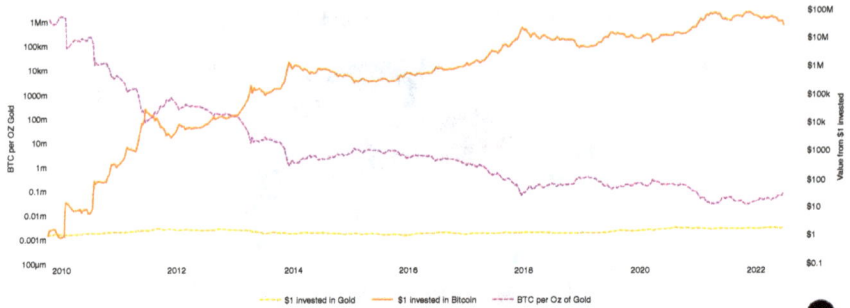

Figure 2.   Bitcoin vs Gold over 12.7 years (Bitcoin: $26,331,500; Gold: $1.77).
Source: Woobull Charts.[3] This chart tracks Gold vs Bitcoin performance from a $1 investment on 6 Oct 2009 when Bitcon first had a market price. I humbly dedicated this chart to @PeterSchiff in honour of his tireless promotion of Bitcoin to his audience of gold bugs, we are forever grateful.

portfolios (Strack, 2021; Yahoo Finance, 2022). In recent months, however, gold is picking up again.

Although Bitcoin has long maintained a low correlation with the US stock market, it has increasingly moved in tandem with Nasdaq, especially the large-cap technology stocks (see Figure 3). This relationship implies that Bitcoin, as an asset category, increasingly shows more characteristics of a technology stock than those of a "mine".

1.3.   *Bitcoin vs crypto*

Bitcoin represents the core value of blockchain.

In terms of market capitalization, Bitcoin usually takes over 40% of the overall market share. During bull markets, the market share shifts to other cryptocurrencies. When the market turns bearish, Bitcoin's market share increases. This phenomenon has created only not the perception that Bitcoin can be the "ultimate" collateral for the cryptocurrency assets class but also a proposition that PoS is viable because PoW works.

In terms of blockchain network design, the PoW network architecture and verification mechanism are no longer mainstream in

---

[3]Woobull Charts (2022). https://charts.woobull.com/bitcoin-vs-gold/.

Even Closer
Bitcoin's correlation with the Nasdaq 100 has reached a new high

Correlation(NDX Index,PR005,40,0) (XBTUSD BGN) 0.6945

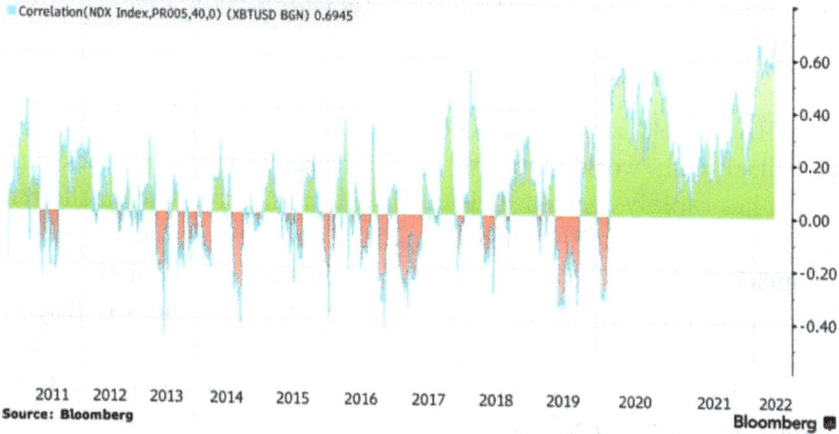

Figure 3.   Bitcoin's correlation with Nasdaq 100.
Source: Ossinger (2022). Bloomberg, https://www.bloomberg.com/news/articles/
2022-04-11/bitcoin-s-correlation-with-big-tech-increases-to-record-chart.

developing new blockchains. However, Bitcoin has affirmed PoW's core values after several rounds of hard forks. That is, ultimate security and value storage. Lightning Network at Layer 2 handles Bitcoin payments. Smart contracts functionality is mostly run on the Ethereum blockchain and other Layer 1 chains and communicated to Bitcoin via cross-chain bridges or centralized exchanges.

The belated Taproot upgrade in November 2021 brought a new level of security, privacy and scalability to Bitcoin (CYBAVO, 2021). A mainstream application is yet to emerge, but the upgrade has brought much excitement to potential projects.

1.4.   *Bitcoin vs decentralized autonomous organization (DAO)*

In addition to providing one of the most trusted native assets in the cryptocurrency world, the significance of Bitcoin for Web3 may lie in a newer approach to organization. It showcases the successful collaboration between human and machine, or between human and human, through code-based governance. It proves that a job that requires massive worldwide collaboration can be completely trustless.

1.5. *Bitcoin vs the world*

The Bitcoin blockchain has been referred to as the cornerstone of the blockchain world. On top of this cornerstone, the blockchain world has rapidly evolved. It is getting deeply entrenched in the real world, impacting entities such as financial institutions, regulators, the financially excluded crowd, and players in the tech industry. Their participation in the blockchain world, in turn, has molded Bitcoin into a different shape.

The Bitcoin blockchain is a bridge connecting the virtual and real worlds, and somehow, WAGMI (We're All Gonna Make It).

2. Ethereum: A Smart Contract Platform

Since the introduction of Bitcoin in 2009, smart contracts by Ethereum have been hailed as the most innovative invention. The Ethereum is a public blockchain platform with smart contract functionality, which allows anyone to build decentralized applications. It lays a solid foundation for the emergence of decentralized applications (DApps) and fuels the decentralized finance (DeFi) and non-fungible token (NFT) boom afterwards.

2.1. *Smart contracts*

A smart contract is a programmable contract or a self-executing program. A critical prerequisite for a smart contract to bear any value is a tamper-proof storage and execution layer so that it cannot be corrupted or modified.

Turing-complete smart contracts leapfrog Bitcoin from the limitation of being a simple ledger and open the door to complex value transfer. Smart contracts fit blockchain naturally due to the tamper-proof nature of blockchain; it also extends the usage of blockchain technology from a sole payment system to more complicated use cases. What follows is a multitude of use cases that demand greater blockchain performance. The surge in demand paves the way for

Table 1. Major public chains and the contract languages.

Name of the chain	Contract language	Turing completeness
Bitcoin	Scripting language	No
Ether	Solidity	Yes
BNB (BSC)	Solidity	Yes
Avalanche	Solidity	Yes
Solana	Rust	Yes
Cosmos	Rust\Go	Yes
PolkaDot	Rust	Yes
Fantom	Solidity	Yes
Polygon	Solidity	Yes
Flow	Cadence	Yes
Dfinity	Motoko	Yes
Near	Rust	Yes

Figure 4. Percentage of locked-in value for different smart contract languages. *Source*: The Block (2022), theblock.co.

introducing high-performance public chains and Layer 2 projects. Table 1 shows major public chains and the contract languages.

At present, Ethereum remains the largest smart contract platform. Its contract language, Solidity, is the most popular programming language. Solidity accounts for 85% of the total value locked (TVL) in the DeFi application ecosystem (see Figure 4).

Ethereum-decentralized applications are concentrated in the DeFi field, including decentralized exchange (e.g., Uniswap), decentralized lending (e.g., Aave, Compound), derivatives (e.g., dYdX), and stablecoins (e.g., MakerDAO, Frax). Other applications are mainly in the fields of NFT and GameFi.

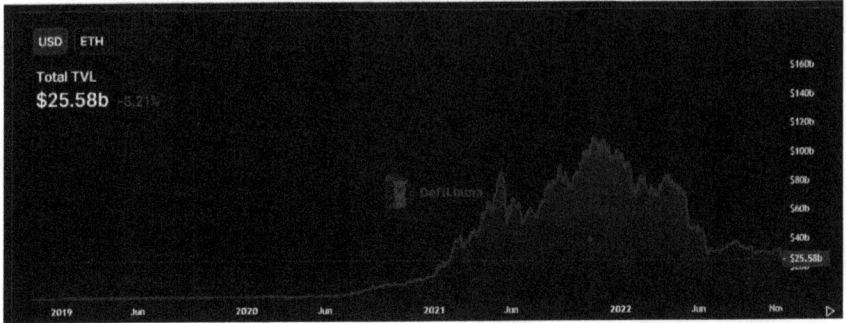

Figure 5.   Ethereum TVL.
Source: DeFiLlama (DeFiLlama, 2022a).

Currently, the TVL on Ethereum is \$25.58 billion (see Figure 5). Figure 5 shows a history of total TVL on Ethereum, which peaked at \$100 billion at late 2021. June 2022 saw stabilization at around \$47 billion after a market fall, which was comparable to the market capitalization of some established technology firms, such as MediaTek (\$35.52 billion) and Kuaishou Technology (\$38.19 billion). The top three applications in the ecosystem are MakerDAO, Lido, and Uniswap, accounting for 16.7%, 10.3%, and 9.9% of Ethereum's TVL, respectively.

2.2. *Ethereum and EVM-compatible chains*

Compatibility with the Ethereum Virtual Machine (EVM) is necessary for many public chains and Layer 2 solutions today.

As the public chain with the largest ecosystem and developers, Ethereum plays a pivotal role in the public chain community. Hundreds of public chains and EVM-compatible chains are active on the market, but very few manage to form their own moats surrounding their ecosystem. All public chains have shifted their focus on improving transaction per second (TPS) to a two-wheel drive growth model stressing ecosystem building and cryptoeconomic incentivization.

Ethereum has always been a leading success in developing its own ecosystem. It is increasingly irreplaceable owing to the gradual advancement in the Ethereum Merge[4] and sharding. Various public chains are proactive in becoming EVM-compatible so their developers can quickly implement DApp migration and deployment. For instance:

BNB Smart Chain (BSC): BNB Smart Chain (BSC) was launched by the cryptocurrency exchange Binance on September 1, 2020.[5] As the first EVM-compatible public chain launched in the "DeFi summer", BSC received most of the traffic from the Binance platform and rose to prominence in the public chain field. BSC adopted a Delegated Proof-of-Stake (DPoS) mechanism similar to EOS.[6] Its transactions per second (TPS) is 30–70 times faster than the Ethereum blockchain. However, it is limited to only 21 validators. Hence, its degree of decentralization is much lesser compared with Ethereum.

Avalanche-C: Avalanche is an interoperable and highly scalable decentralized public chain network.[7] Avalanche is divided into X-chain, C-chain, and P-chain. Among them, the C-chain is an EVM-compatible chain with smart contract functionality. The X-chain is a Directed Acyclic Graph (DAG) structure with the fastest transaction speed and is mainly used for money transfers. The P-chain is the metadata blockchain that coordinates validators, keeps track of active subnets, and enables the creation of new subnets. The P-chain is mainly used for staking, similar to the Relay Chain of the Polkadot network.

Fantom: Fantom is a high-performance public chain based on DAG technology and supports EVM compatibility.[8] Backed by Andre Cronje, an influential and prolific developer nicknamed the "father of DeFi", Fantom has experienced explosive growth in its ecosystem development in the past year. However, with Andre

---

[4]At the time of writing, the Ethereum Merge has completed on September 15, 2022.
[5]BNBCHAIN.ORG. https://www.bnbchain.org/en.
[6]EOS.IO. https://eos.io/.
[7]AVAX.NETWORK. https://www.avax.network/.
[8]FANTOM.FOUNDATION. https://fantom.foundation/.

Cronje's announcement to withdraw from the DeFi circle at the beginning of 2022, Fantom has also seen its darkest moment. Its TVL dropped from a high of $11.81 billion to $980 million, equating to a drop of 91.7% according to the website DeFiLlama.

Nearly all mainstream public chains are compatible with EVM. In addition to the three public chains, many public chains that were initially EVM-incompatible have launched second layers that are compatible with Ethereum. For example, near launched Aurora; PolkaDot launched Moonbeam; Evmos on Cosmos; and Neon on Solana are also EVM-compatible. This effort will further strengthen the influence of Ethereum in this field.

2.3. The Ethereum merge: From PoW to PoS

As one of the core components underlying blockchain, the consensus mechanism is the golden standard to maintain state consistency in the blockchain network. It determines the attribution and distribution of bookkeeping rights. Different versions of verification mechanisms have evolved in public chains. But in terms of applications, two major mechanisms gain the most traction, namely PoW and PoS.

PoW is best represented by Bitcoin, while the new generations of public chains, such as BSC, Fantom, and Ethereum 2.0, use PoS. In the PoS mechanism, validators no longer spend immense computing power to compete for bookkeeping rights. Validators need only to create and submit blocks when they are randomly selected or get rewarded by verifying blocks submitted by others when they are not selected.

"The Merge" refers to the merger of the Ethereum Mainnet and the beacon chain. The Merge is an essential step in the transition from the Ethereum Mainnet to the era of sharding (Ethereum Foundation, 2022b). According to the Ethereum Foundation, the consensus layer is merged with the execution layer. The consensus layer refers to the beacon chain, and the execution layer refers to the existing layer that interacts with Ethereum. After the Merge, Ethereum will abandon the PoW part of the current execution layer and fully shift to PoS. At that time, the Ethereum network will be validated by stakers, and the PoW mining nodes will be retired.

The low scalability, high energy consumption, and high gas fees of the Ethereum network have seriously restricted the development of its ecosystem. Sharding is the optimal solution to solve the above problems. The implementation of sharding has become a priority for Ethereum's future development. The Merge lays the foundation for developers to focus on further developing sharding.

In fact, the transition from PoW to PoS has been mapped in the development roadmap of Ethereum. Its design of the "difficulty bomb" is a special mechanism that incentivizes PoW miners for the transition to PoS. The "difficulty bomb" is a mechanism to adjust the mining difficulty according to the block time. The algorithm and the increase in block height will show an exponential increase in block difficulty. The result is that, at a certain point, miners will not be profitable after weighing the mining cost. They would quit and move from PoW to PoS. Due to rounds of delays in the Merge, the difficulty bomb was postponed several times. The Grey Glacier Hard Fork in June 2022 also signaled that the Ethereum Merge would not happen until at least after September 2022 (Redman, 2022).

There are three main changes to be brought about by The Merge. Firstly, the creation of Ether is to be significantly reduced. Under the PoW mechanism, about 12,000 Ethers were minted per day. After switching to PoS, there will only be about 1,280 Ethers per day, resulting in a decrease of 89.3%. In addition, with the burning mechanism of EIP-1559, Ether is likely to deflate.

Secondly, the barrier of entry to validate blocks is lowered, which is conducive to further decentralization of the network. Under the PoW mechanism, miners need professional mining machines. This barrier makes it difficult for ordinary users to become a miner on the network. But under the PoS mechanism, block validators no longer compete on computing power. The requirement for hardware significantly lowered. Anyone can run their node as a validator after meeting the staking requirement. The emergence of various staking service providers has further lowered the barrier to becoming an Ethereum validator.

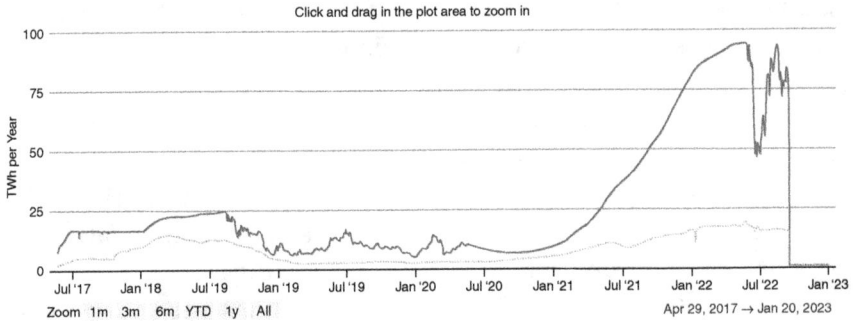

Figure 6. Ethereum energy consumption index (the solid line: estimated TWh per year; the dotted line: minimum TWh per year).
Source: Digiconomist (2022), https://digiconomist.net/ethereum-energy-con sumption.

Finally, the energy consumption due to mining will be greatly reduced, and we will gradually move toward the era of carbon neutrality. The PoS mechanism eliminates the continuous pursuit of high computing power mining machines and significantly reduces electricity consumption. Currently, the annual energy consumption of the Ethereum network is about 51.32 TWh, which is as high as the energy consumption of Portugal as a nation (49TWh). The annual carbon dioxide emission reaches 28.63 tons. According to the Ethereum Foundation calculations, the energy consumption will be reduced by 99.95% after the Merge (see Figure 6). The energy consumption of each node per day will be comparable to that of a personal computer for home use (Ethereum Foundation, 2022).

It is important to note that The Merge is insufficient to bring about scalability and gas fee improvements. The gradual implementa- tion of sharding will manifest significant changes in these two aspects.

3. Layer 2 on Ethereum

To improve the performance of the Ethereum network, various scaling solutions have been proposed in the industry. Depending on the consensus levels involved, these solutions can be categorized into Layer 1 or Layer 2 solutions.

Layer 1 is on-chain scaling, which usually achieves performance improvement by increasing the block size or the blockchain's underlying data structure. Sharding falls under Layer 1 scaling. Sharding is divided into transaction sharding and state sharding. Transaction sharding refers to mapping the data processing into different shards according to certain rules. State sharding pertains to storing the data separately according to the different attributes of shards and improving network performance through parallel processing in different shards.

Layer 2 refers to off-chain scaling, which migrates data computation, transactions, and other services to another layer outside the main chain to reduce the workload on the main chain. Various Layer 2 scaling solutions have emerged to ensure the usability and security of the Layer 2 data, such as ZK Rollup, Optimistic Rollup, Validium, Plasma, etc.

Before the eventual implementation of sharding in Ethereum, Layer 2 solution remains the best scaling solution for Ethereum. Currently, the Ethereum Layer 2 solution is mainly based on two main rollup solutions, zero-knowledge rollup (ZK Rollup), and optimistic rollup.

Rollup, as the name suggests, means transaction aggregation. Transactions are processed separately, aggregated, and submitted to the main chain at one time to reduce the frequency of interaction with the main chain and achieve the goal of reducing network congestion and improving network performance. Rollup ensures that the original transaction data is stored on the Ethereum main chain without relying on specific verification nodes.

3.1. *ZK Rollup*

ZK Rollup was first introduced in 2018. It relies on zero-knowledge proofs to secure funds. It allows proof that one is the rightful owner of certain rights without disclosing identity-relevant information. The Ethereum main chain serves as the storage medium and is used to verify the final state, so ZK Rollup also inherits the security properties of the main chain.

ZK Rollup protects users' funds from confiscation and censorship, but the technical immaturity and hurdle to building an interoperable network restrict the solution's applicability. For ZK Rollup adopters, creating an interoperable EVM-compatible environment is much more difficult than Optimistic Rollup. The most representative projects of ZK Rollup are zkSync and StarkNet.

zkSync: zkSync is developed by the Matter Labs team and is fully EVM-compatible with the 2.0 test network online.[9] In zkSync 2.0, the Layer 2 state is divided into zkRollup with on-chain data availability and zkPorter with off-chain data availability, similar to StarkNet and StarkEx under StarkWare. There are nearly 100 officially announced on-chain projects focusing on infrastructure, cross-chain bridges, and DeFi. In the zkSync network, the gas fees can be paid by other tokens, not necessarily by Ether.

StarkNet: StarkNet is a Layer 2 scaling platform by Stark-Ware.[10] Although it belongs to the same ZK Rollup family as zkSync, the solution is slightly different. StarkNet uses zk-SNARKs, which requires less on-chain storage space and gas fees, while zkSync uses zk-STARKs, which is better in network security.

StarkNet financed $100 million at a valuation of $8 billion in May 2022, making it one of the highest valued among all Layer 2 projects (Bambysheva, 2022). StarkWare is actively testing its L1–L2 bridge, Starkgate, on their website. Seventy projects are displayed on the official website, mainly in the DeFi domain.

3.2. *Optimistic Rollup*

Optimistic Rollup adopts fraud proof, instead of zero-knowledge proof. It draws on the early Plasma scaling technology that relies on the game between the verifying node and the challenger to protect fund security. Therefore, when the verifying node returns the final

---

[9]ZKSYNC.IO. https://zksync.io/.
[10]STARKNET.IO. https://starknet.io/.

state of the transaction data on Layer 2 to the main chain, it will enter a challenge window of about seven days, and the funds will be locked during the challenge period. If the verified transaction data is proven faulty, other validators can submit fraud proofs and will get the locked fund from the original verifying node.

Compared to ZK Rollup, Optimistic Rollup has the significant advantage of being compatible with more complex smart contracts. Thus, existing Layer 2 projects that are launched have adopted Optimistic Rollup. Some examples are given below.

Optimism: Optimism is the first to develop an EVM-compatible Optimistic Rollup solution.[11] It guarantees the validity of data synchronized to Layer 1 through a single-round-interactive-proving scheme, which is the main difference with Arbitrum. It is the first to issue tokens among the four major Layer 2 projects.

Arbitrum: Arbitrum, developed by OffChainLabs team, was born at Princeton University.[12] Currently, it is the most complete in terms of ecosystem. Arbitrum uses multi-round interactive fraud proving. After a verifier submits the fraud proof, Arbitrum will first narrow down the scope of the dispute through multi-round interactions at Layer 2 before going to the main chain for simulation. This practice reduces the cost to resolve dispute on chain. The multi-round dispute resolution is the main difference between Arbitrum and the Optimism solution.

3.3. *Validium and Plasma*

Validium (StarkEx): Validium is a hybrid scaling approach led by StarkWare, a zero-knowledge proof research and development organization.[13] It is very similar to the ZK Rollup solution. A key difference is that Validium's transaction data is not stored on the main chain like ZK Rollup's. The proof of validity is published on chain, but the data is stored off-chain. Hence, it is not as secure

---

[11]OPTIMISM.IO. https://www.optimism.io/.
[12]ARBITRUM.IO. https://arbitrum.io/.
[13]VALIDIUM.COM. https://validium.com/.

as the ZK Rollup solution. For example, the StarkEx Validium operator can freeze users' funds.

In addition, it offers limited support for general-purpose computing and smart contracts. The high computing power required to generate zero-knowledge proofs is not cost-effective for low-throughput applications. So far, its advantages are mainly in the absence of withdrawal delay and very high throughput (TPS of about 10,000). Some representative projects applying this solution are Immutable X[14] and DeversiFi.[15,16]

> **Plasma:** In 2017, Plasma was the mainstream Ethereum scaling solution. It is an early scaling technology. To date, along with the maturity of Rollup solutions, Plasma gradually phases out as a Layer 2 solution of less security.

Plasma borrows the technology from the Bitcoin Lightning Network. It has a separate blockchain anchored to the Ethereum main chain and uses proof of fraud to arbitrate disputes. It has the advantage of high throughput and low cost per transaction. But the disadvantages are obvious. For example, it lacks support for general-purpose computing and only supports a few transaction types, such as basic pass-through transfers, exchanges, and certain types of transactions. It also requires regular monitoring or delegating others to monitor the network to ensure funds' security. The most representative Plasma scaling solution is the OMG Network.[17]

In summary, the differences of the Layer 2 solutions lies in the trade-offs between security and scalability (see Table 2).

4. Avalanche: Avalanche Protocol, EVM, Subnet

Avalanche focuses on high performance and high scalability. High performance is achieved by the design of the Avalanche protocol,

---

[14]Immutable X Whitepaper. (https://assets.website-files.com/62535c6262b90af d768b9b26/6304335ed396fd9c8d8dfe5e_Immutable%20X%20Whitepaper.pdf.

[15]DeversiFi was changed to Rhino. Fi since July 2022, according to its website.

[16]RHINO.FI. https://rhino.fi/.

[17]OMG.NETWORK. https://omg.network/.

Table 2. The differences between the four Layer 2 solutions.

Layer 2 solution	Proof mechanism	Security	Scalability	Representative projects
ZK rollup	Zero knowledge proof	High	Low	zkSync StarkNet
Optimistic rollup	Fraud proof	Medium	High	Optimism Arbitrum Metis
Validium	Zero knowledge proof	Low	Medium	Immutable DeversiFi
Plasma	Fraud proof	Low	Low	OMG Network

while customized subnets achieve high scalability. Avalanche is highly compatible with EVM, and this characteristic attracts port-in of well-developed protocols from the Ethereum ecosystem. At the same time, it also facilitates developers in building Avalanche's native protocols.

4.1. *Avalanche protocols*

As the name suggests (Team Rocket, 2018), the consensus process of the Avalanche protocol starts with a random collapse (i.e., statistical results from random sampling) and ends with a large collapse (i.e., consensus formation). Its core idea is that by continuously and repeatedly sampling nodes in the network and collecting their responses to a certain proposal, it can eventually direct all honest nodes to reach a consensus.

The advantages of the Avalanche protocol include high performance, low latency, resistance to Byzantine attacks, resistance to double-spending attacks, and no conflict between the miners and users interests.

Its possible problems are as follows:

- Random sampling may result in a non-deterministic consensus.
- There is no resolution for disputable transactions.
- It requires the support of a large number of participants.

(see ipfs.io/ipfs/QmUy4jh5mGNZvLkjies1RWM4YuvJh5o2FYopN PVYwrRVGV for details).

4.2. Avalanche's network design and native cross-chain bridge

The Avalanche features three built-in blockchains, as shown in Figure 7.

1. The X-Chain (Exchange Chain) is responsible for asset creation and transaction.
2. The P-Chain (Platform Chain) is responsible for storing data on the chain, coordinating validators, and creating new subnets.
3. The C-Chain (Contract Chain) is responsible for executing smart contracts and supporting EVM.

Avalanche Bridge is the native cross-chain bridge to support relaying assets from Ethereum to Avalanche. Additionally, to enable the use of BTC in the Avalanche DeFi ecosystem, a native cross-chain bridge for BTC is added (Ava Labs, 2022).

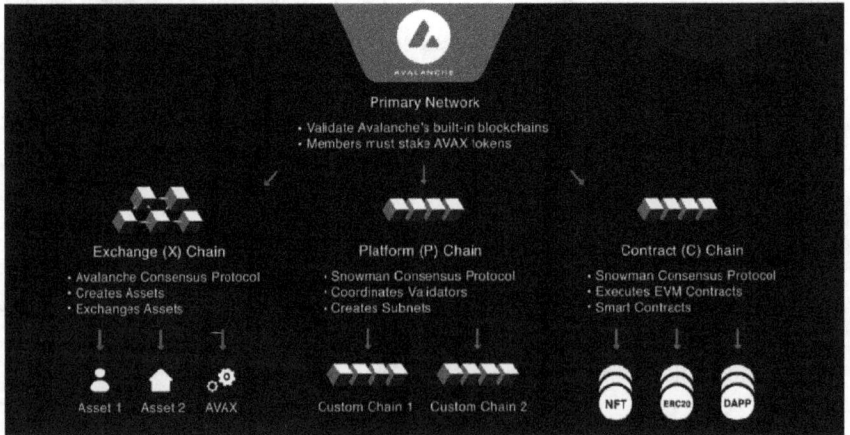

Figure 7. Avalanche platform.
Source: Avalanche website Avax.network.

4.3. *Ecosystem*

Avalanche's high compatibility with the Ethereum ecosystem has not only attracted a large number of Ethereum-native projects but also facilitated many Avalanche-native protocols. Users simply need to add the Avalanche-C chain using the MetaMask wallet to participate in the Avalanche ecosystem.[18]

Currently, the TVL on Avalanche amounts to $2.8 billion. The top five DApps are:

- **Aave**: an Ethereum-native lending protocol with cross-chain deployment to Avalanche
- **Trader Joe**: an Avalanche-native decentralized exchange
- **Wonderland**: an Avalanche-native DeFi 2.0 protocol, that is, the OlympusDAO fork
- **Benqi**: an Avalanche-native lending protocol
- **Platypus Finance**: an Avalanche-native stablecoin exchange

Other featured native protocols are:

- Avalaunch (the largest Launchpad on Avalanche)
- Crabada (once the most active GameFi protocol on Avalanche)
- Yeti Finance (lending protocol on Avalanche, with support for leverage)
- Yield Yak (revenue aggregator on Avalanche)
- Step.app (Move to Earn (M2E) project on Avalanche)
- Ascenders (RPG-type GameFi project on Avalanche)

4.4. *Subnet*

Avalanche supports the deployment of DApps to the Avalanche subnets to build their multi-chain applications.

Subnets are easy to deploy and are EVM-compatible. Currently, subnets cannot communicate directly with each other. The first project deployed in the Avalanche subnet is the DeFi Kingdom, followed by deployment plans for Crabada, Step.app, and Ascenders (for a full list of projects, see www.subnet.tech).

---

[18]METAMASK.IO. https://metamask.io/.

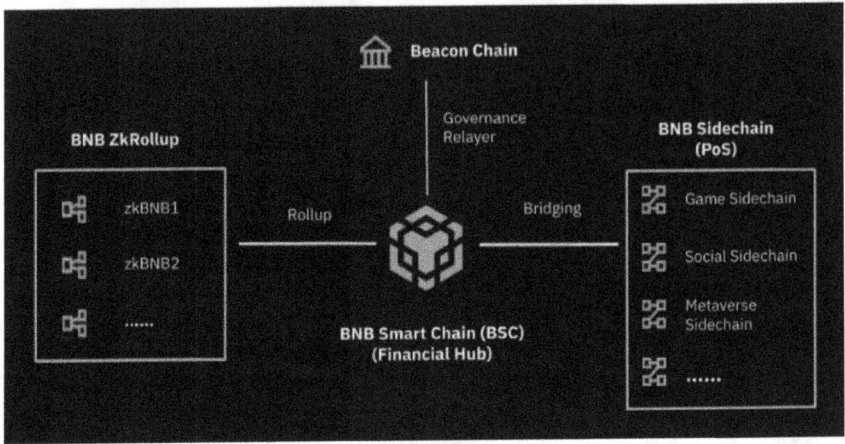

Figure 8. BNB network design.
Source: Binance blog.[19]

5. BNB Chain: Binance, EVM, BAS

The Build and Build (BNB) Chain is closely linked to Binance — the world's largest centralized exchange. It is EVM-compatible and its sidechain is called the BNB Application Sidechain (BAS) (bnbchain.org, 2022).

5.1. *Architecture*

As shown in Figure 8, the BNB Beacon Chain is responsible for the governance of the BNB chain (e.g., staking, voting). The BNB Smart Chain (BSC) is EVM-compatible; it has a consensus layer and it is a hub connecting multiple chains. In addition, the BNB sidechain uses existing BSC functionalities to develop PoS solution for customized blockchain and DApps. The upcoming BNB ZkRollup plans to extend the BSC into a hyper-performing blockchain whereas the BSC Partition Chain (BPC) offloads some computations from the BNB Beacon Chain similar to the Ethereum's Layer-2 solution (Binance Academy, 2020).

---

[19]Binance blog. Binance.com/en/blog.

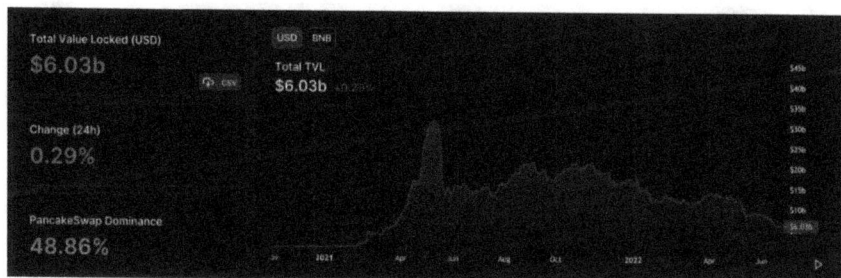

Figure 9. Total value locked for BSC chain.
Source: DefiLlama, https://defillama.com/chain/BSC (DeFiLlama, 2022b).

5.2. *The Binance coin (BNB)*

Unlike the native tokens of other public chains, the Binance coin (BNB) is not only the main coin of the BSC but also the platform token of the Binance cryptocurrency exchange. BNB is not only affected by activities on the BSC but also closely linked to the trading volume and business revenue of the cryptocurrency exchange.

BNB passed the proposed BEP-95 real-time burning mechanism in November 2021 (BnB Chain Developer, 2021). The approved amount of BNB burnt in the proposal is not conducive to complex smart contract interactions for GameFi type of projects in the long term; it may result in a significant increase in the ease-of-use threshold for similar applications. Given the BAS establishment, it is speculated that, in the future, the BSC will host high-frequency smart contract interactions primarily on the sidechain.

5.3. *Ecosystem*

Figure 9 shows that the TVL on the BSC is now about $6 billion, accounting for 7.8% of the TVL on all the chains (DeFiLlama, 2022b).

Among the projects in the ecosystem, PancakeSwap accounts for 48.86% of the total TVL. Nearly all of the top ten TVL projects are native to the BSC (see Figure 10), seven of which have been listed on the Binance Exchange.

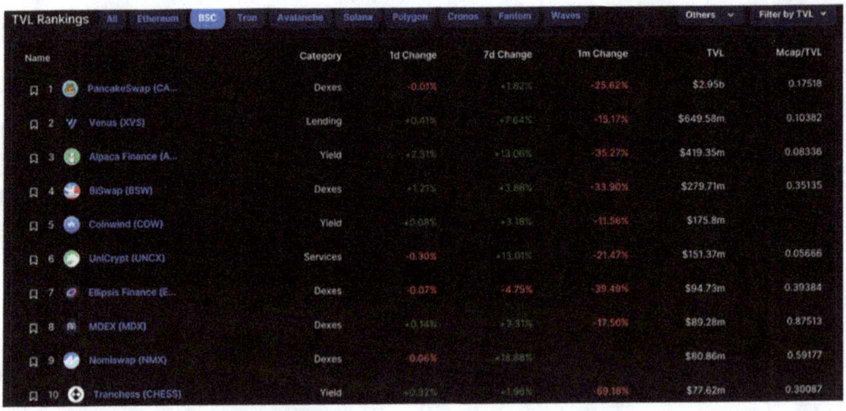

TVL Rankings All Ethereum BSC Tron Avalanche Solana Polygon Cronos Fantom Waves Others Filter by TVL

Name	Category	1d Change	7d Change	1m Change	TVL	Mcap/TVL
1 PancakeSwap (CA...	Dexes	-0.01%	+1.82%	-25.62%	$2.95b	0.17518
2 Venus (XVS)	Lending	+0.41%	+7.64%	-15.17%	$649.58m	0.10382
3 Alpaca Finance (A...	Yield	+2.31%	+13.06%	-35.27%	$419.35m	0.08336
4 BiSwap (BSW)	Dexes	+1.21%	+3.88%	-33.90%	$279.71m	0.35135
5 Coinwind (COW)	Yield	+0.08%	+3.18%	-11.56%	$175.8m	
6 UniCrypt (UNCX)	Services	-0.30%	+13.01%	-21.47%	$151.37m	0.05666
7 Ellipsis Finance (E...	Dexes	-0.07%	-4.75%	-39.49%	$94.73m	0.39384
8 MDEX (MDX)	Dexes	+0.14%	+3.31%	-17.50%	$89.28m	0.87513
9 Nomiswap (NMX)	Dexes	0.06%	+18.88%		$80.86m	0.59177
10 Tranchess (CHESS)	Yield	+0.32%	+1.96%	-69.18%	$77.62m	0.30087

Figure 10. Top 10 projects on BSC chain according to TVL.
Source: DefiLlama, https://defillama.com/chain/BSC (DeFiLlama, 2022b)

Thanks to the relatively low development cost on BSC, many projects are active on BSC. Hash daily transaction once reached 16 million in November 2021 (see Figure 11).[20]

The BSC has a large number of active DeFi projects (e.g., Tranchess, GameFi projects (e.g., Binary X), and metaverse projects (e.g., SecondLive).[21] The only thing missing is an NFT trading platform, although mature trading platforms are already on the market.

BSC has provided generous support for project development, aiming to enrich the ecosystem. It regularly hosts its Most Valuable Builder (MVB) programs to select and incubate outstanding projects. As of October 2021, a $1 billion fund was pledged to support promising BSC projects.[22]

5.4. *The BAS*

According to (Mehta, 2022) each BAS will have — three to seven verifiers of its own and is expected to run a PoS-based super majority

---

[20]BSCSCAN.COM. https://bscscan.com/chart/tx.
[21]BnbProject.com. https://bnbproject.org/#/.
[22]BNBCHAIN.ORG. (2021). https://www.bnbchain.org/en/blog/2021-a-ground breaking-year-for-bsc/.

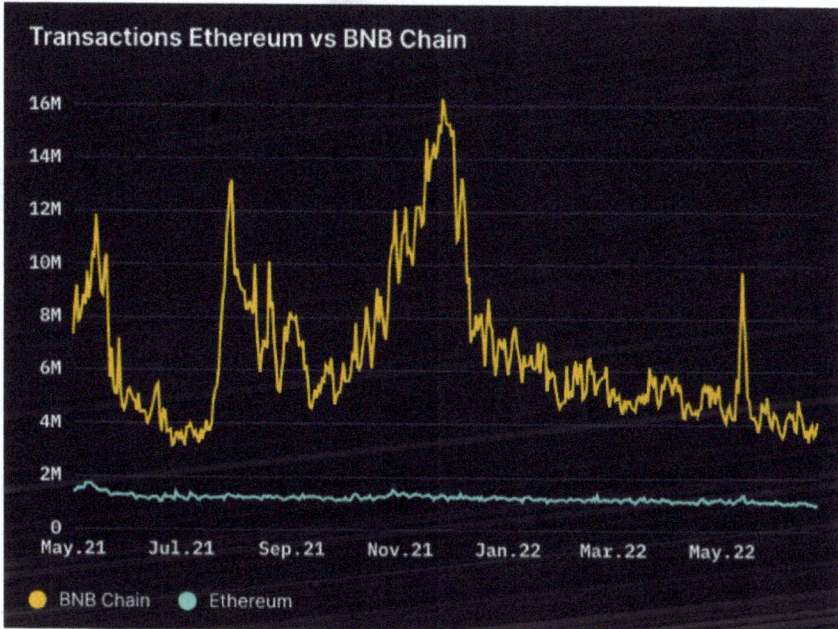

Figure 11.   Transaction on Ethereum vs BSB Chain.
Source: defiprime.com.

(2/3) consensus. Each BAS will operate using its stake and tokens. In addition, each sidechain's state and state transitions will be completely independent of the other sidechains.

BAS chains will require third-party bridges to communicate with each other. In this case, the BSC will use Celer's third-party bridge connect to each BAS via a "lock + mint" format (Stakingbits, 2021), while each BAS is also connected via this mechanism.

The current confirmed BAS projects include Meta Apes (a BSC chain-native GameFi), Project Galaxy (an on-chain identity credential project for multi-chain deployments), and Cube (a BSC chain-native gaming platform).[23]

---

[23]baschainlist.com. https://baschainlist.com/.

6. Cosmos: Open Architecture, Modularity, and Airdrops

As a founding project of the multi-chain architecture, "openness" is the word that describes the philosophy of the Cosmos ecosystem.

6.1. *Open architecture: Shared security and interchain accounts*

Figure 12 shows a diagram of the Cosmos architecture. As shown in the figure, the core of the Cosmos architecture is the Tendermint consensus engine. Any application chain can theoretically call upon

Figure 12.   Cosmos architecture diagram.
Source: X Consulting.[24]

---

[24]X Consulting. https://0xzx.com/20190516205675959.html.

this encapsulated consensus generation module through the Application Blockchain Interface (ACBI). (*Note*: the ABCI is the green pillar that connects Tendermint to the Cosmos Hub in Figure 12.)

There are two kinds of chains in the upper layer: Hub Chains as the main "router" to relay and Zone Chains to run applications. The two chains communicate through the Inter-Blockchain Communication (IBC) protocol. The cross-chain communication is further strengthened by interchain accounts, which can be used to complete operations across different chains in one go.

In theory, such architecture allows each Zone to be connected to Tendermint via ABCI and form a completely independent chain. However, independence also implies autonomy. The chain is vulnerable to attacks if there are not enough stakers. Hence, after the official launch of the first Hub (i.e., the Cosmos Hub), many Zones opt to connect directly to it and share the security brought by the large number of $ATOM stakers on the Cosmos Hub, and then indirectly connect to all other Zones in the ecosystem through the Cosmos Hub. As a result, Cosmos becomes an entity with shared security.

6.2. *Modularization of development tools for cosmos SDK*

The Cosmos SDK toolkit, which is packaged into modules, is the most user-friendly development tool for blockchain application developers (See Figure 13). By deploying commonly used modules, developers can quickly assemble the generic modules of their application and focus on their specialized modules. At the same time, the SDK also standardizes and encapsulates the latest and most commonly used modules for developers to avoid duplicating development efforts.

6.3. *Airdrops*

Due to the shared security, the validation of newly added application chains is largely processed by other chains. In return for this contribution, new projects typically airdrop their tokens to $ATOM and several other major chains (e.g., Osmosis, Juno, Secret).

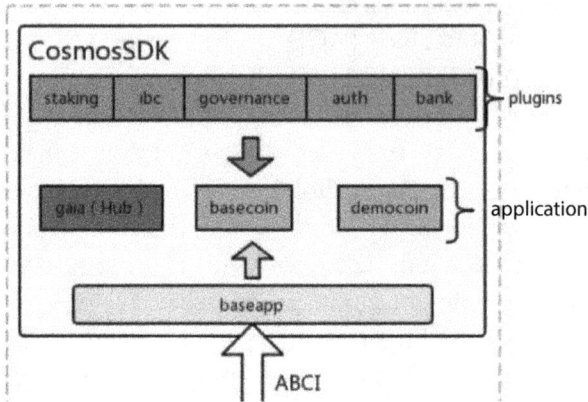

Figure 13. Modules for Cosmos DSK.
Source: Hui Ge (2019), cloud.tencent.com/developer/article/1446970.

The frequent airdrops have also led to another unintended consequence — experimentations and reflections on the decentralized autonomous organization (DAO) airdrop mechanism and the improvement in governance issues that come with it.

Several major airdrops include Osmosis (took place on July 4, 2021); Juno (took place on August 27, 2021); and Evmos (took place on April 19, 2022). Among them, the Juno airdrop sparked a major controversy about DAO governance.

Cosmos labels itself "the Internet of blockchains". It is characterized by its openness, modularity, and airdrops. To many, this model has the potential to become the underlying Layer 0 of all blockchains.

7. PolkaDot: Relay Chain, Parallel Chain, Slot Auction, and Hackathon

PolkaDot aims to build a network that can transmit all data across all blockchains and its current development is converging with other Layer 1 solutions; it is also focusing more on substantiating its ecosystem.

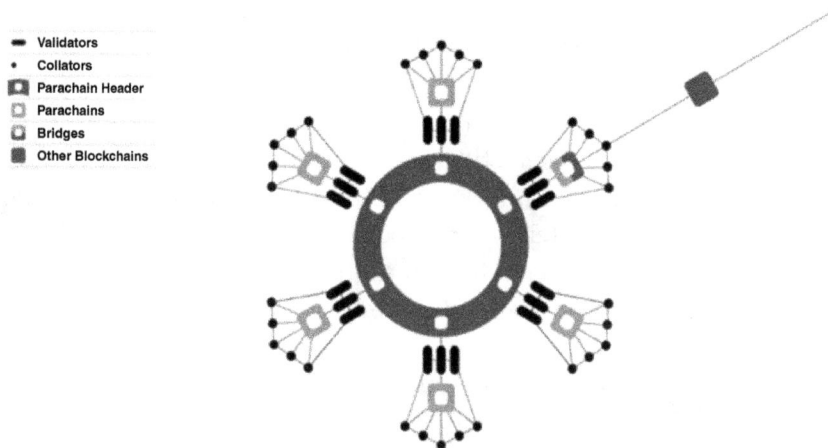

Figure 14. PolkaDot architecture.
Source: Hui Ge (2019), cloud.tencent.com/developer/article/1446970.

7.1. *Architecture: Relay chain and parallel chain*

PolkaDot architecture, as shown in Figure 14, features a multi-chain network. In PolkaDot's multi-chain network (Wood, 2017), all the chains are either a relay chain or a parallel chain. Relay chains provide the underlying PoS verification, shared computation, and consensus. Parallel chains run different applications and are connected to relay chains by slots. Other chains that are not parallel chains (such as the Ethereum and the Bitcoin blockchains) can communicate with the relay chain via bridges, a special type of parallel chain that specializes in cross-chain communication.

(For technical details, refer to the Polka Whitepaper by Wood (2017): polkadot.network/PolkaDotPaper.pdf).

7.2. *Slot auction*

To use the relay chain and join the PolkaDot ecosystem, project participants need to bid for slot positions (capped at about 100) for a two-year lease period. The successful bidders will have their $DOT locked during this period.

The first round of slot auction on December 2021 has attracted a stake of 99,113,200 \$DOT (8.6% of the total). Five successful bidders are Acala Network, Moonbeam Network, Astar Network, Parallel Finance, and Clover Finance (Polkadot, 2021).

Six projects won the auction in the second round with 27 million \$DOT (2.4% of the total), including Efinity, Centrifuge, Composable Finance, HydraDX, Interlay and Nodle.

As PolkaDot has only limited slots, other networks, such as Kusama, the canary network for PolkaDot, are also continuously auctioning slots.

Unlike block records that eventually reach a state of finality, the blockchain state is dynamic and constantly evolving. PolkaDot's development direction has transited from "the ruler of cross-chain" to Layer 0 development and then to Layer 1-like development. This transition, to some extent, reflects the shift in design thinking of blockchain public chains.

8.   Solana: PoH, Ecosystem, Downtime Events

Among all mainstream public chains, Solana is perhaps the most unique. The special Proof of History (PoH) consensus mechanism, the adoption of Rust, the complete lower layer ecosystem for DeFi and NFT, and the network's "delightful" distributed denial of service (DDOS) attacks contribute to Solana's uniqueness. In terms of the design concept, Solana looks like a "counterattack" by "out-of-the-loop" programmers against their blockchain counterparts.

8.1.   *Mechanisms: Rust, PoH, and "triangular trade-offs"*

Rust is not a mainstream blockchain programming language, as most chains use EVM's Solidity system. But in the 2020 survey of "Stack Overflow" (Lewkowicz, 2020), Rust was the "most popular programming language," with about 86% of developers stating they wanted to continue coding in Rust in the future (Supra Oracles, 2022).

In a meeting between Solana, Zcash, and Parity on September 23, 2018, the Solana founders summarized six reasons why Rust is a good fit for blockchain development: (1) it is as fast as C/C++; (2) it is type-safe as Haskell; (3) there is no garbage collector, and variables are automatically recycled out of scope to free memory; (4) it eliminates null pointers and hanging pointers, which are the root cause of C/C++ system crashes and unsafe code; (5) strict rules; (6) it supports concurrent programming (Solana, 2018).

Solana's PoH consensus mechanism uses a very innovative asynchronous structure. Generally speaking, blockchain requires network-wide synchronization when the state is updated. In other words, all nodes are synchronized to update before the next block is produced. And this reduces the node's efficiency. Solana introduces a piecewise clock and a global clock to maximize the performance of each node. As such, state updates no longer require global time synchronization, and each node will periodically synchronize its clock with the global clock.

Solana also introduces a verifiable delay function (VDF) to address the trust issue of transactions. PoH records a timestamp when each transaction is packaged on the chain so that nodes can use VDF to verify the history of operations on the chain.

The efficient Rust language and the PoH consensus of running nodes under full load are the pillars of the "hyper-speedy" Solana. Bitcoin and Ethereum main chains have sacrificed scalability in the blockchain scalability trilemma, while Solana has sacrificed decentralization.

Currently, the Solana Foundation is the only entity developing core nodes on the blockchain. According to Solana beach (https://solanabeach.io/), the current number of nodes in Solana is 1,793, with a Satoshi Nakamoto factor of 26 (Nakamoto factor: the minimum number of entities needed to compromise a given subsystem). So theoretically, only 26 nodes are needed to bring Solana down.

8.2. *Ecosystem: Serum and Metaplex*

According to the Solana official website, as of June 25 2022, there are 301 DeFi projects on Solana (including 175 DEX, 25 AMM mechanisms, and 150 Order Book mechanisms), 929 NFT projects (including 100 Metaplex-related ones), and 271 games. The DeFi system is based on Serum, while the NFT system is based on Metaplex.

Half of the DeFi projects are DEX, thanks to the DeFi infrastructure Serum on Solana.

Serum is an Order Book DEX where all DEX liquidity on Solana is pooled into Serum. In other words, if there is a pending order on any DEX, it is Serum that ultimately aggregates the trade, and your counterparty is all the Maker of the DEX on Solana. All DEX is just a GUI for Serum and this ensures that the liquidity on Solana is concentrated.

In addition, Solana's close relationship with the centralized exchange FTX allows Serum to share some of the off-chain liquidity.

Compared to the DeFi projects, Solana has twice as many NFT projects. Metaplex, the underlying NFT protocol on Solana, supports users in a one-stop process of casting, pricing, and offers. Solana is probably the best public chain for NFT at the moment.

In the era where everything can be an NFT, it dramatically lowers the threshold for NFT creation. With a well-articulated vision and a good story, NFTs can hit the market. Therefore, when the market of Ethereum NFT is cold, the NFT on Solana is heated.

When the market was in the doldrums in May 2022, the OpenSea (Ethereum chain) trading volume fell by 31.6% year-on-year. While Magic Eden (the largest NFT trading platform on the Solana chain) saw a 39.79% year-to-year increase in trading volume, OpenSea (Solana chain) also saw a 286.02% year-to-year increase.

8.3. *Downtime*

Although Solana mainly promotes its high TPS and fast transaction processing speed, it is often very unstable. Below are some of the more severe incidents on the main chain:

- On May 1, 2022, the Solana Mainnet was flooded with 4 million requests per second, causing nodes to run out of memory and stop issuing blocks for nearly seven hours.
- On May 26, 2022, a block clock mismatch occurred on the Solana Mainnet, where the on-chain timing was about 30 minutes behind the real-world timing.
- On June 1, 2022, Solana's Mainnet was down for about 4.5 hours as blocks failed to reach consensus.

There were also several times where Solana Mainnet suffered "performance degradations".[25]

The reason for the constant downtime is that new blockchain games, NFT minting activities, and Genesis NFT sales attracted a large number of scientists and their bots. Solana was continuously subjected to at least double-digit per second hits by the various bots, subjecting the network to DDOS attacks (a large number of invalid requests that prevented normal requests from being submitted). For example, the downtime on May 1, 2022 was also caused by a bot attack on Candy Machine, a tool for NFT minting introduced by Metaplex. The hit on StepN also caused congestion in Solana.

Since then, Solana has been imposing a penalty of 0.01 $Sol every time a wallet submits an invalid NFT transaction.

---

[25](See https://status.solana.com/history for details, node updates Twitter: @SolanaStatus.)

The main sources of Solana's problems are twofold: the underlying technology and the NFT boom. Solana may have been able to resist the traffic brought by arbitrage bots for the DeFi liquidation, but it lost out to the NFT bots.

If speedy synchronization is Solana's trademark, then downtime is the cost it pays. But compared to 2021, Solana's performance is gradually improving. For example, TPS is recovering, and transactional failures are declining. Perhaps these are "growing pains," as Solana Labs founder Anatoly Yakovenko put it.

9. Blockchain in China: Digital Collections + Consortium Blockchain

After the regulatory changes in 2021 (Zhang, 2021), blockchain in China is mainly dominated by NFT platforms and consortium chains with a limited number of nodes controlled by a handful of developers. The top 100 platforms in China are backed by several big enterprises (Gao, 2022) (see Table 3).

In addition, there are Bilibili and Bigverse (NFT China), which both use ETH to issue NFT and other companies using Solana and Polygon.

In view of decentralization, consortium blockchains are subject to controversy. The failure of Meta's (formerly Facebook) Libra project may have exemplified the failure of the consortium chain. But it is too early to conclude that there is no room for consortium chain in Web3.

The history of blockchain can almost be equated to the history of public chains. What is reflected in different iterations of public chains are different understandings of the current world by different communities and varied approaches to tackling different problems. Old solutions will eventually become new problems. One thing that is certain about the future of Web3 is that the public chain will be the core underlying technology for a long time and it will continue to evolve.

Table 3. The top blockchain platforms in China

Consortium chain	Developers	Platforms
Ant Chain	Ali	Whale Quest, Phantom Tibet, Super Dimensional Space, Seven Level Universe, Blue Cat Digital, One Flower yihua, Ninth Space, Mysterious Oasis, Digital Traveler DT Universe
BSN Consortium Chain	Wenchang Chain (based on IRITA); Taian Chain (based on FISCO BCOS); Wuhan Chain (based on Ethereum); Tangshan Chain (based on DBChain); Guangyuan Chain (based on Everscale); Zhongyi Chain (based on EOS)	Mytrol Digital Wenchuang; Tiger Yuan Yuan, Digital Collection China, He Luo, Qian Xun Shu Cang
Zhixin Chain	Tencent	Phantom Core, TME Digital Collection, R-Digital Collection
TUSI Chain	Tencent	Yunxui Digital Collection
Baidu Super Chain	Baidu	Baidu Super Chain, DongYiYuanDian, LiangShi Digital, YuanShi, LinQian
Starfire Chain	Ministry of Industry and Information Technology	Zebra China, Ali Auction
Tianhe Chain	Tianhe State Cloud	You Copyright, Museum Chain
Juntou Chain	Juntou.com	Hi Universe
Jingdong Zhi Zhen Chain	Jingdong	Ling Rare, Red Fruit Number Collection
Interesting Chain	Interesting Chain Technology	Red hole collection, law core
Netease Blockchain	Netease	Netease Planet
Hespi Chain	Mi Chain Technology	Grass Square
Yuanjing Union Chain	Hangzhou Yuanjing Technology	West Lake One

Note: The English translations were based on Chinese spelling, when a platform's official English names were not identified at the time of the manuscript preparation.
Source: Guo Zhihao, https://www.163.com/dy/article/H6PEPLNQ0552DGBG.html.

References

Ava Labs, I. (2022). Avalanche platform. https://docs.avax.network/overview/getting-started/avalanche-platform.

Bambysheva, N. (2022). Ethereum scaling company StarkWare quadruples valuation to $8 Billion amid bear market. *Forbes*. https://www.forbes.com/sites/ninabambysheva/2022/05/25/ethereum-scaling-company-starkware-quadruples-valuation-to-8-billion-amid-bear-market/?sh=35c496226699.

Binance Academy. (2020). BNB smart chain: A parallel blockchain to beacon chain to enable smart contracts. https://github.com/bnb-chain/whitepaper/blob/master/WHITEPAPER.md#design-principles.

BnB Chain Developer. (2021). Introducing BEP-95 with a real-time burning mechanism. https://www.bnbchain.org/en/blog/introducing-bep-95-with-a-real-time-burning-mechanism/.

BNBCHAIN.ORG. (2022). Getting started with BNB sidechain. https://docs.bnbchain.org/docs/BNBSidechain/overview/bs-overview/.

CYBAVO. (2021). Taproot: Bitcoin's future-proofing privacy and scalability upgrade. https://www.cybavo.com/blog/bitcoin-taproot-upgrade/.

DeFiLlama. (2022a). Etherum TVL.

DeFiLlama. (2022b). Total value locked for BSC chain. https://defillama.com/chain/BSC.

Digiconomist.net. (2022). Ethereum energy consumption index. https://digiconomist.net/ethereum-energy-consumption.

Ethereum Foundation. (2022). The Merge. https://ethereum.org/en/upgrades/merge/#:~:text=The%20upgrade%20from%20the%20original,energy%20consumption%20by%20 99.95%25.

Gao, Z. (2022). https://www.163.com/dy/article/H6PEPLNQ0552DGBG.html.

Hetzber, C. (2022). El Salvador's millennial president launching Bitcoin 'volcano bond' in major bet on cryptocurrency craze. *Fortune*. https://fortune.com/2022/03/14/el-salvador-president-bitcoin-city-volcano-bond-nayib-bukele/.

Hui Ge. (2019). cloud.tencent.com/developer/article/1446970.

Jenkinson, G. (2022, July 6). El Salvador's bitcoin wallet chivo scores $52M in remittances in 2022. *Cointelegraph*. https://cointelegraph.com/news/el-salvador-s-bitcoin-wallet-chivo-scores-52m-in-remittances-in-2022.

Kedem, S. (2022, July 22). Central African Republic freezes adoption of bitcoin. *African Business*. https://african.business/2022/07/technology-information/what-you-need-to-know-about-the-central-african-republics-adoption-of-bitcoin/.

Lavie, D. (2022). Bitcoin halving is how the supply of the world's largest cryptocurrency is controlled. https://www.businessinsider.com/personal-finance/bitcoin-halvin.

Lewkowicz, J. (2020). Report: Rust is the most beloved programming language for five years running. *SD Times*. https://sdtimes.com/softwaredev/report-rust-is-the-most-beloved-programming-language-for-five-years-running/.

McDonald, M. (2022). The infuriating reality of traveling with bitcoin in the world's crypto capital. *Bloomberg*. https://www.bloomberg.com/features/2022-bitcoin-travel-problems/?leadSource=uverify%20wall.

Mehta, S. K. (2022). Flavors of standalone multichain architecture. Jump crpto.com. https://jumpcrypto.com/flavors-of-standalone-multichain-archi tecture-2/.

Nakamoto, S. (2008). Bitcoin: A peer-to-peer electronic cash system.

Ossinger, J. (2022). Bitcoin's correlation with big tech increases to record. *Bloomberg*.

Polkadot. (2021). Making history, again: Polkadot auctions 1-5. https:// polkadot.network/blog/making-history-again-polkadot-auctions-1-5/.

Redman, J. (2022). Ethereum devs delay difficulty bomb -ETH 2.0 contract surpasses ether deposits. https://news.bitcoin.com/ethereum-devs-delay-difficulty-bomb-eth-2-0-contract-surpasses-13-million-ether-deposits/.

Renteria, N., Wilson, T., & Strohecker, K. (2021, June 10). In a world first, El Salvador makes bitcoin legal tender. *Reuters*. https://www.reuters. com/world/americas/el-salvador-approves-first-law-bitcoin-legal-tender-2021-06-09/.

Solana. (2018). https://medium.com/solana-labs/solana-at-portland-dev-meet up-72e4dc7ad32c.

Stakingbits. (2021). How to bridge tokens cross-chain using Celer cBridge. https://medium.com/stakingbits/how-to-bridge-tokens-cross-chain-using-celer-cbridge-b14d26340a1c.

Strack, B. (2021). Ray Dalio: Crypto should be part of a diversified portfolio. Blockworks.co. https://blockworks.co/ray-dalio-crypto-should-be-part-of-a-diversified-portfolio/.

Supra Oracles. (2022). Blockchain programming languages for ambitious web3 developers. https://supraoracles.com/academy/gaining-an-edge-in-web3-development-solidity-haskell-go-rust/.

Team Rocket. (2018). Snowflake to avalanche: A novel metastable consensus protocol family for cryptocurrencies. https://ipfs.io/ipfs/QmUy4jh5mGNZ vLkjies1RWM4YuvJh5o2FYopNPVYwrRVGV.

The Bank for International Settlements. (2022). The future monetary system. *BIS Annual Economic Report Issue*. https://www.bis.org/publ/arpdf/ar202 2e3.htm.

The Block. (2022). Percentage of locked-in value for different smart contract languages. theblock.co.

Waters, K. & Maddrey, N. (2022). Halfway to the next halving: The state of bitcoin in 8 charts. https://newsletterest.com/message/99306/Coin-Metrics-State-of-the-Network-Issue-151.

Wood, G. (2017). Polkadot: Vision for a heterogeneous multi-chain framework. polkadot.network/PolkaDotPaper.pdf.

Yahoo Finance. (2022). Yahoo finance presents: Ray Dalio. https://sg.finance. yahoo.com/video/yahoo-finance-presents-ray-dalio-120000362.html.

Zhang, Z. (2021). China 2021 recap: Overview of a pivotal year. *China Briefing*. https://www.china-briefing.com/news/china-2021-recap-key-developments-impacting-business-foreign-investors/.

Chapter 3

Infrastructure — *Extension*

Mozhi Wang, Xueyan Qin, Swee Won Lo, and Yu-chen Hung

The public chain is still considered a minimum viable product (MVP) in the blockchain landscape. Even after Ethereum started the era of smart contracts, there are still many other "components" needed to extend the functionality of public chains to support direct applications so that they can be easily used by developers, researchers, and general users.

1. Identity System: Private Key "Translator"

From a traditional point of view, the identity system in Web3 is account-based. Technically, it is everything that a string of private key can correspond to on a distributed ledger.

In Web3, where accounts (addresses) are generated irreversibly from passwords (private keys), private keys constitute the lowest layer of identity. All Web3 phenomena revolve around the private key. (*Note*: The public key is derived from the private key through elliptic curve cryptography, and the address is derived from the public key through hashing.)

But the problem is that the underlying "key" is a hash (Damgård, 1990). So, the identity system (project) has to "translate" this hash. This conversion is the most important infrastructure besides the design of the public chain itself.

1.1. *Identity system v1.0: Mnemonic and wallet*

Private keys and mnemonic phrases have a one-to-one correspondence. Compared with the hash value, the mnemonic phrase can be remembered easier, albeit they are still (a string of) words with no coherent meaning. Wallet applications will use mnemonic as the first layer of simplified translation of the private key.

The Ethereum blockchain, layer-1 EVM chains using the same structure as Ethereum (e.g., AMAX,[1] BNB chain (Huang *et al.*, n.d.), Axie,[2] etc.), and Ethereum's layer-2 solutions (e.g., Polygon (Kanani *et al.*, n.d.), Arbitrum,[3] Optimism,[4]) can all share a standard wallet. Each chain tends to develop its wallet application due to its different underlying authentication methods and private key formats. This is the reason that the MetaMask wallet is also the most widely used of all wallet applications.

In addition, Phantom is available on Solana[5]; Keplr on Cosmos,[6] and Polkadot JS on Polkadot.[7]

1.2. *Identity system v2.0: Domains and privacy*

Version 2.0 of the identity system is primarily an enhancement of addresses. Addresses are meant to be visible to everyone, and there are two requirements for addresses: easy to remember (e.g., for acquaintance transfers) and privacy (e.g., for high-value transactions).

---

[1] Armonia Meta Chain. (n.d.). Retrieved February 12, 2023, from https://amax.network/#/.

[2] Axie Infinity. (n.d.). Retrieved February 12, 2023, from https://whitepaper.axieinfinity.com/.

[3] Arbitrum — Scaling Ethereum. (n.d.). Retrieved May 1, 2023, from https://arbitrum.io/.

[4] Optimism. (n.d.). Retrieved February 12, 2023, from https://www.optimism.io/.

[5] Phantom Technologies, Inc. (n.d.). Phantom — A friendly Solana wallet built for DeFi & NFTs. Phantom. Retrieved February 12, 2023, from https://phantom.app/.

[6] Documentation|Keplr Wallet. (n.d.). Retrieved May 1, 2023, from https://docs.keplr.app.

[7] Polkadot{.js}. (n.d.). Retrieved February 12, 2023, from https://polkadot.js.org/.

Addresses are made memorable by purchasing a domain name non-fungible token (NFT), each of which may correspond to only one address at a time. For example, Ethereum's Ethereum Name Service (ENS),[8] Solana's Bonfida,[9] and MYNFT on Flow.[10]

In terms of privacy, addresses are pseudonymous rather than anonymous. The incorporation of privacy techniques makes addresses anonymous. The prime example is the Monero Coin (The Monero Project, n.d.), which bakes privacy features into the blockchain.

1.3. *Identity system v3.0: KYC, soul tokens, electronic signatures, multisignatures*

The identity system of the current 3.0 era goes further in terms of the correspondence between addresses and private keys.

There are already projects that bind social accounts to addresses (Know Your Customer). For example, domain name projects that can add Twitter accounts and BrightID (BrightID Main LLC, n.d.) require a user's real identity to participate in a Zoom meeting. In addition, Rabbithole — a type of soul-binding NFT (Buterin, 2022) that is non-transferable — can also be considered an innovation that branches out from on-chain identity.

Mnemonic phrases are not necessarily the only avenue to recover private keys. In principle, any information corresponding to a unique private key can be used. This information can be biological information (fingerprint, voice, face recognition) and legal documents (business licenses, attestations). There are no mature projects yet, but this development direction has become a trend.

Multi-signature wallets are currently used mainly by institutions. They are modeled after a voting system similar to a board of directors' approach, where transactions can only be executed with the signed consent of the majority of the board members controlling

---

[8]ENS (n.d.). Ethereum Name Service. Ethereum Name Service. Retrieved February 12, 2023, from https://ens.domains.

[9]Solana Name Service | Bonfida (n.d.). Solana Name Service |Bonfida. Retrieved February 12, 2023, from https://naming.bonfida.org/.

[10]MYNFT.IO (n.d.). Mynft. Retrieved February 12, 2023, from https://mynft.io/.

the wallet. On-chain asset inheritance in the future may bring the need for multi-signature to every crypto asset holder (Di Nicola *et al.*, 2020; Goldston *et al.*, 2023).

2. Cross-Chain Bridges

Each public chain (such as BSC, Polkadot, Avalanche, and Fantom) has formed its ecological map in the multi-chain era. Along with the booming development of DeFi, NFT, and GameFi, enabling cross-chain for assets is becoming an immediate need. Cross-chain bridges (Kannengießer *et al.*, 2020) garnered attention for several reasons — the pursuit of profits on different public chains or layer-2 chains for on-chain assets, the pursuit of good user experience for different on-chain applications, and the developers' strategy for multi-chain applications.

As a bridging tool, cross-chain bridges remove the ecological silos, realize the transfer and portability of assets between different chains, and promote blockchain interoperability (Pillai *et al.*, 2022).

In summary, the significance of cross-chain bridges is mainly reflected in the following points.

- The emergence of cross-chain bridges allows on-chain assets to be easily transferred between chains, significantly improving the utilization rate of on-chain assets. The improvement of asset utilization will, in turn, promote the development of on-chain applications.
- Cross-chain bridges provide a better user experience, allowing users to conveniently utilize different applications across multiple networks. They reduce operational barriers and open many possibilities for utilizing applications on different chains.
- Cross-chain bridges can form the foundation layer of multi-chain protocols. With the rapid development of public chain networks, it is difficult for a single blockchain system to handle the demand of various applications, and blockchain interoperability has become increasingly important. Through cross-chain bridges, various dApps can provide richer services and open up to more user groups to enhance their competitiveness.

2.1. *Cross-chain model*

At present, there are three main cross-chain models.

Lock+Mint: Lock-and-mint is one of the most common methods that enable cross-chain asset transfer. In this process, an asset is locked in a specific smart contract at the source chain, and an asset of equal value is minted in the other chain. This asset is essentially a redeemable credential. Lock-and-mint cross-chain is often used when introducing mainstream assets to new chains, so it is generally viewed as an official bridge. An example is the Avalanche Bridge.

Two-Way Liquidity Pool Model: The two-way liquidity pool model is commonly used in third-party cross-chain bridges. The cross-chain bridge pre-sets up a fixed pool of assets on the source and destination chains. As users use the cross-chain bridge, they transfer the asset to the smart contract address on the source chain and receive the corresponding cross-chain asset from the destination chain. Since the two-way liquidity pool generally requires officially issued cross-chain assets, it can only be set up after the official cross-chain bridge is established.

Atomic Swap: Atomic swap is the most decentralized cross-chain model, as it is achieved by using self-executing smart contracts.

From the validator's perspective, most cross-chain bridges currently use a third-party validation model, where one or more third parties will verify cross-chain information. The third party (or parties) will monitor and verify the information on the source chain. Once the validators reach a consensus, the corresponding assets will be minted on the destination chain.

Some cross-chain bridges also require validators to stake a certain amount of assets for security and to prevent validators from committing fraud. Depending on the number of validators, the validation model can be either multi- or single-point. In addition, techniques such as multi-signature or multi-party computation (MPC) are also used in multi-point validation to enhance security (see Table 1).

As the development of cross-chain bridges advances, each cross-chain bridge will have advantages for different chains, resulting in

Table 1. Similarities and differences of some cross-chain models.

Name	Authentication mechanism	Asset cross-chain model	Multi- or single-chain bridge
Hop	Multi-point validation and asset staking	Liquidity Pool	Multi
Multi-chain	Multi-point validation and MPC	Lock+Mint	Multi
Rainbow	Light client and relay validation	Lock+Mint	Single
cBridge	Multi-point validation, multi-signature and asset staking	Atomic Swap	Multi
Stargate	Oracle and relay	Liquidity Pool	Multi

different cross-chain costs and cross-chain speeds. However, it is counter-intuitive to be requiring users to navigate around different cross-chain bridges. Hence, cross-chain aggregators are designed to automatically match the most suitable cross-chain links by aggregating existing cross-chains. Li.Finance[11] and XY Finance (XY FINANCE, 2023) are examples of cross-chain aggregators.

In addition, unique "public chains" such as Cosmos and Boka have their multi-chain ecology that is taking shape. The working principle relies on the underlying cross-chain IBC and relay chains of their respective chains (Ou *et al.*, 2022). Unlike the cross-chain bridges mentioned above, which mostly rely on third parties, it realizes cross-chain assets and greatly facilitates cross-chain communication. Both the Cosmos and Boka's cross-chain are atomic-level cross-chains. Using the light client and relay verification method, they have high security and they enabled real asset cross-chain instead of asset value cross-chain.

2.2. *Cross-chain bridge security*

From the $600 million theft from the Axie Ronin bridge (Thurman, 2022) to the theft of $100 million from Horizon (Browne, 2022),

---

[11]LI.FI Bridge Aggregator (n.d.). LI.FI. Retrieved February 12, 2023, from https://li.fi/.

the official cross-chain bridge of Harmony, the security issue of cross-chain bridge has become an important factor limiting its development. It has triggered a collective lack of user confidence in cross-chain bridges. The main groups affected by the current security issue of cross-chain bridges are the verifiers and the liquidity providers.

The common cross-chain bridge vulnerabilities mainly occur in two places: one is at the cross-chain signature, where attackers obtain the public and private key pairs of project parties or verifiers by various means to steal assets; the second is to directly exploit contract vulnerabilities, such as the previous Wormhole cross-chain bridge theft incident, where the attackers deceived multiple signatures and minted 120,000 Wormhole ETH through contract vulnerabilities (Chainalysis, 2022), stealing a large number of assets.

3. Distributed Storage: Mechanisms, Tracks, Issues

As one of Web3's fundamental infrastructures, distributed storage is necessary to decentralize large-scale on-chain data, addressing issues such as single point of failure and data losses. Compared to centralized storage, decentralized storage is also highly competitive in terms of privacy protection, data security, and speed (see Table 2).

Filecoin(IPFS): The InterPlanetary File System (IPFS) protocol (Benet, 2014) was developed by Juan Benet, who later founded Protocol Labs. As a new transport protocol, IPFS is benchmarked

Table 2.   Centralized vs distributed storage.

Storage type	Centralized storage	Decentralized storage
Data security	Weak	Strong
Privacy protection	Weak	Strong
Download speed	Slow	Fast
Idle storage utilization	Low	High
Storage incentive	None	High
Storage cost	Low	High

Source: Zahed Benisi *et al.* (2020).

against the Hypertext Transfer Protocol (HTTP) (HTTP Documentation, n.d.), commonly used in the traditional Internet. Filecoin (Filecoin, n.d.) belongs to the incentive layer of IPFS, aiming to solve the IPFS network stability and security issues and attract professional storage service providers to provide professional and secure storage services through the incentive mechanism.

Filecoin miners can be categorized as storage miners or retrieval miners, responsible for storing and retrieving files, respectively. These miners charge fees to the user. Currently, Filecoin is actively promoting the development of the Filecoin Virtual Machine (FVM),[12] which is compatible with the Ethereum Virtual Machine (EVM). The FVM introduces smart contracts on the existing storage layer to achieve intelligent and programmable storage, allowing Filecoin to take a differentiated competitive route in the current storage track.

Arweave: Arweave (Arweave — A Community-Driven Ecosystem, n.d.) is a permanent data storage protocol. The Arweave network encourages miners to save as many blocks as possible using Succinct Proofs of Random Access (SPoRA). It first randomly selects a Recall Block, and only miners who have saved this Recall Block can participate in the mining of the subsequent blocks. By this mechanism, it realizes large-scale replication of the data, achieving permanent storage.

At present, the demand for large-scale storage is not strong yet due to the performance limitations of public chains. Hence, the demand on storage space is still not high. However, the issue of data redundancy becomes highly significant, especially when facing problems such as data explosion.

Storj: The Storj network (Storj Labs Inc., n.d.) mainly includes three roles: users, satellites, and nodes. Users put forward the requests of uploading or downloading data, nodes are responsible for executing the requests, and satellites are responsible for choosing the fastest nodes, recording data such as users' and nodes' expenditure and

---

[12]Filecoin Virtual Machine (n.d.). Retrieved February 12, 2023, from https://fvm.filecoin.io.

income, storage location, and nodes' reputation. According to Storj, it mainly anchors Amazon S3's decentralized cloud storage platform, so its token has no complex mechanism or economic model. As a project founded in 2014, Storj is far less promising than Filecoin and Arweave in terms of development prospects and existing market applications.

The role of distributed storage for Web3 is not yet prominent. The data stored are mainly NFT images, on-chain contracts, on-chain graphics, and other data that occupy little memory and have low interactivity. However, as the metaverse concept develops rapidly, the demand for large-volume data storage for video, audio, and digital model will bring new development opportunities for distributed storage.

Distributed storage also faces many practical issues, similar to on-chain finance.

Firstly, there is the regulatory problem of illegal content. Blockchain's tamper-resistant nature results in an uncontrollable spread of unlawful content. This issue is the main factor impeding distributed storage. No matter what ideal sentiment blockchain geeks have, the system has to operate within society's legal and moral framework. Therefore, constructing an end-to-end regulatory model starting from data in-flow to out-flow is a problem necessary for distributed storage solutions to go mainstream.

The second issue is to improve the storage space utilization rate. Current distributed data storage has a large portion of junk and redundancy. Once large-scale applications become ubiquitous, it remains to be seen whether the current distributed storage projects can quickly adapt to the extremely rapid changes in the market.

Lastly, at present, each distributed storage project has not demonstrated the storage's durability, which is also an area to be verified.

4. Oracle: Mechanisms, Types, and Application Scenarios

Blockchain cannot obtain off-chain data because it cannot establish external connections on its own; this "inability" secures the blockchain compared to traditional networks. However, blockchain must use an intermediate layer that manages data to realize actual and practical application scenarios. The intermediate layer is the oracle.

4.1. *Mechanism*

The blockchain oracle connects the off-chain world and the on-chain world. It allows the blockchain to connect to the payment system, data providers, exchanges, cloud providers, Internet-of-Things (IoT) devices, and other off-chain environments.

According to Chainlink (Chainlink, n.d.), an oracle typically possesses the following key functions.

- Listen
- Extract
- Format
- Validate
- Compute
- Broadcast
- Output (option)

(see https://chain.link/education-hub/oracle-problem for details).

4.2. *Oracle types*

According to the IOSG Ventures classification (Gu, 2022), oracles can be divided into four different types (see Table 3).

At the time of writing, Chainlink's decentralized oracle is the main solution used in the market. Despite the efficiency and authority issues, decentralized oracles remove the single point of valuation problem that the other three types of oracles exhibit. In addition, as shown in Figure 1, an ideal Decentralized Oracle Network (DON) could encompass all four data access methods, according to the Chainlink 2.0 whitepaper (Breidenbach *et al.*, 2021). This type of oracle is perhaps the ultimate form.

Table 3. Four different types of oracles according to Gu (2022).

Software Oracle	Using API interfaces provided by third-party servers or websites
Hardware Oracle	Electronic sensor and data collectors in the IoT setting; also includes various barcode scanners, bank card Point-of-Sales machines, medical devices that collect medical data
Centralized Oracle	Generally provided by third parties such as governments or credit agencies
Decentralized Oracle	Aggregation-based processing: Aggregates multiple data sources to eliminate impact of single malicious data (e.g., Chainlink (n.d.)
	Staking-based processing: Requires participants to stake assets to increase credibility, e.g., Band (Band Protocol — Cross-Chain Data Oracle, n.d.)
	Game-theory-based processing: Provides non-adversarial economic incentives, e.g., NEST (NEST Protocol | The Most Important Infrastructure after ETH, n.d.)
	Reputation-based processing: Restricts adversarial nodes using a reputation system, e.g., Witnet (n.d.)

4.3. *Application scenarios*

At the time of writing, oracles are mainly used in asset price-feeding in Web3 applications. However, as Web3 expands into broader domains, application scenarios for oracles are expected to increase.

DeFi: Price Feed and Fair Sequencing: In DeFi, fair sequencing (Juels, 2020) is a new application scenario for oracles in addition to price-feeding.

At present, each transaction is validated and sequenced by miners, and this gives miners great room for arbitrage and manipulation, resulting in a measure called the Miner's Extractable Value (MEV).

Fair sequencing consists of two aspects: first, sorting is performed according to the time that the nodes receive the transactions; second, the causal relationship between the transactions is preserved.

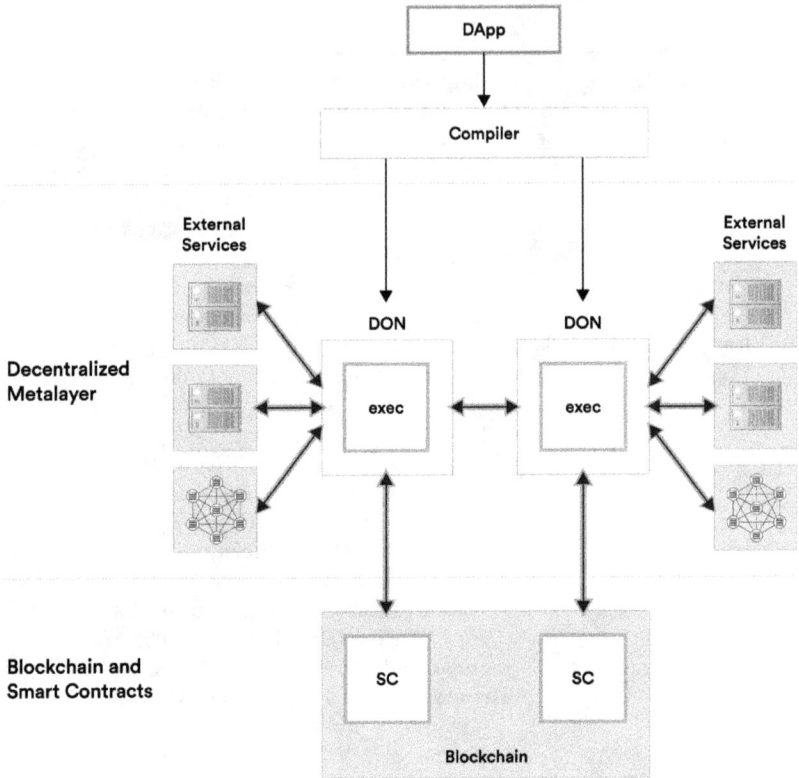

Figure 1. A Decentralized Oracle Network as depicted in Breidenbach *et al.* (2021).

Source: Breidenbach *et al.* (2021).

Causality refers to hiding the transaction data and waiting until the consensus layer finishes sorting before disclosing the order of transactions to prevent changing the causality between transactions. From this, it can be seen that fair sequencing can effectively counter frontrunning.

NFT and GameFi: Random Numbers: Random numbers can be considered a necessity for NFT players and GameFi players. Artists need random numbers to generate NFTs with different rarities randomly; GameFi players also need a fair and just game rule to ensure their rights are not harmed.

It is necessary to ensure that the random numbers are trustworthy. As there are no real random numbers, the oracle's VRF (Verifiable Random Function) (Chainlink, n.d.) generates a series of real and verifiable random numbers off-chain and provides them for use. This process ensures the confidentiality and fairness of the random numbers.

CCIP: Cross-Chain Information: CCIP (Cross-Chain Interoperability Protocol) (Chainlink Foundation, 2023) provides developers with an open-source standard to build secure cross-chain services and applications easily. With a common messaging interface, smart contracts can communicate across multiple blockchain networks, and developers do not need to write code specifically to access a particular blockchain. An example is Chainlink's Messaging Router.

DID: Authentication: DID (Decentralized IDentifier) can be fully utilized in countering Sybil attacks (an attack where one attacker creates multiple account nodes to control the network). DID can be fully utilized in the Decentralized Oracle Network (DON) (Breidenbach *et al.*, 2021).

The oracle has quietly unfolded a new narrative, shifting from its function of price-feeding to a multi-faceted role supporting various emerging applications. The oracle is an essential and integral part of Web3, and it also serves as a good Web3 entry point for traditional industries and government agencies.

5. The "Redemption" of Stablecoin

After the collapse of the Luna coin (Sandor & Genc, 2022), stablecoin became the last straw that broke the camel's back. However, Web3 needs stablecoins as much as the economic system needs general equivalents. While the "BTC standard" has become more of a value proposition, the "U standard", denominated in US dollar stablecoins, is now the norm in Web3, as shown in Figure 2 which compares the trading volume of BTC and USDT.

5.1. *Stablecoin positioning*

Stablecoin is an on-chain token pegged to a fiat currency's value. Stablecoin and bitcoin have a similar niche — they bridge the real

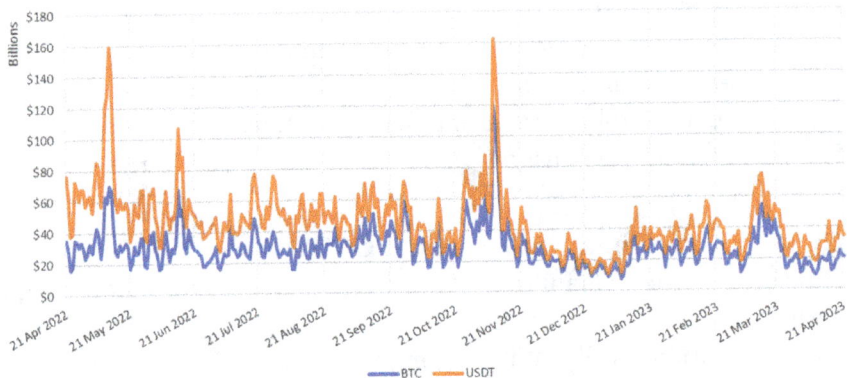

Figure 2.   Daily trading volume — BTC vs USDT.
Source: CoinMarketCap.

and the virtual worlds. In the following, we explore the kind of stablecoin that Web3 needs.

5.2.   *Types of stablecoins*

5.2.1.   On-Chain Stablecoins (Algorithmic Stablecoins and Over-Collateralized Stablecoins)

Numerous examples of algorithmic stablecoins have failed, e.g., Basis (Dale, 2018), ESD (empty set, n.d.), Titan (Adams & Ibert, 2022). Each generation of algorithmic stablecoin issuers has tried to play the role of a "central bank" by regulating the supply to stabilize the price peg. Historically, this has been more difficult than expected (despite mechanisms such as Curve's 3-pool 4-pool, partial collateralization, etc.); even central banks may not be able to achieve this in reality.

Stablecoins created by over-collateralization (e.g., Dai) seems to reduce the risk of the system (an ETH worth $1000 can only generate 800 DAIs, and a drop in the value of ETH would result in liquidation to recover DAIs from the market). But multiple rounds of borrowing and lending increase leverage and amplify system risk.

Stablecoins generated by on-chain assets improve capital utilization more efficiently than other stablecoins in the phase of credit expansion. Hence, it is easy to understand how on-chain stablecoins can grow rapidly during market upturns (e.g., Terra & UST (Sandor & Genc, 2022); Tron & USDD (TRON DAO RESERVE, n.d.); Near & USN (NEAR Foundation, 2022.)). However, as an on-chain credit token that is pegged to the USD, the "death spiral" begins when both on-chain and off-chain liquidity tightens simultaneously.

As an asset that bridges on- and off-chain, convergence at both ends weakens the anti-fragility in the algorithm.

Central Bank Digital Currency (CBDC): A digitalized central bank currency can solve the problem of pegging for stablecoins. However, if CBDCs are adopted in the public chain in a large scale without ensuring proper risk control and mitigation, they can only be used in a limited area on small-scale pilot applications.

In the era where currency digitization and digital transformation take the trend, CBDC has much broader applications in the Web3 world (Lee *et al.*, 2021).

Stablecoins for "Commercial Banks" (USDC vs USDT): Tether (Tether Operations Limited, 2013) and Circle (Circle Internet Financial Limited, 2023) are two "commercial banks" that issue stablecoins. They take in USD savings in real-time and issue 1:1 stablecoins on each chain as "bank certificates of deposit". The chains are responsible for on-chain transactions, and the two "banks" are responsible for off-chain withdrawals and deposits. They purchase low-risk debt in USD and are also regulated by the SEC.

In addition, the USDC (by Circle), which has better compliance, tends to replace the USDT (by Tether) as shown in Figure 3. Meanwhile, Circle announced on July 1, 2022, the launch of Euro Coin, a Euro-backed stablecoin, with the same full-reserve model as USDC (Circle Internet Financial LLC, 2022).

There may be new iterations of the stablecoin in the Web3 economy. To date, all stablecoins have opted for a "commercial bank" stablecoin. Perhaps this is the best way to balance our two worlds.

Figure 4 shows the stablecoin supply (in billion) per day.

Figure 3.   Market Cap of USDC and USDT, respectively.
Source: CoinMarketCap.

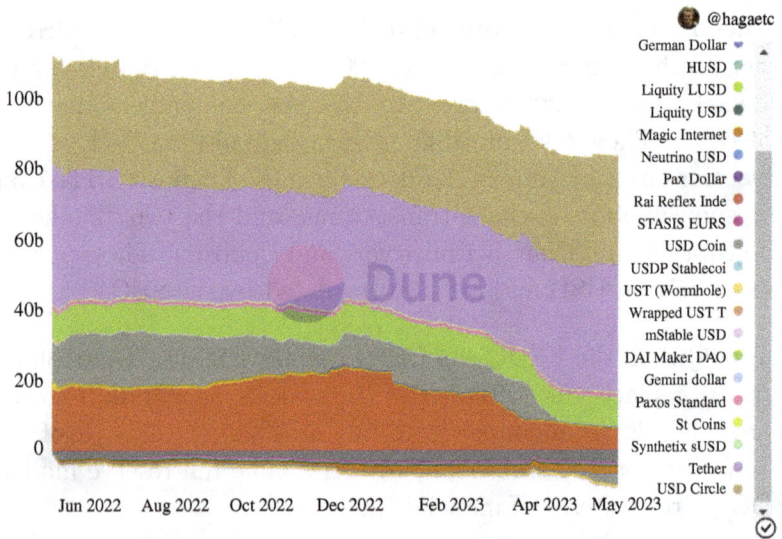

Figure 4.   Stablecoins Supply Per Day Per Type as of May 2023.
Source: Alirezaananas (2020) (as of June 2022).

6. Blockchain Browser: Scan, Index Query, Data Analysis

Every transaction on the blockchain is a form of data. Hence, a blockchain holds a massive amount of data. However, there is a severe lack of functionalities to collect, organize, and search the big data. Projects in this area are still minimal and have a lot of potential for development.

6.1. *Scan*

A blockchain explorer is a tool for users to query all blockchain data. Public chain stakeholders usually develop it, and the architecture is simple and mainly used to display on-chain data, as shown in Figure 5.

6.2. *Indexing*

Google uses web crawlers to crawl through databases to build the existing real-world indexing. However, the nature of the blockchain makes indexing exceptionally difficult to accomplish. Indexing applications primarily represented by The Graph (Graph Foundation, n.d.) attempt to create a "roadmap" for indexing data on the chain (see Figure 6).

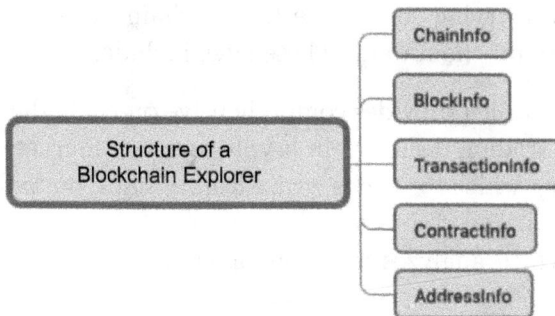

Figure 5. Structure of a Blockchain Explorer.
Source: Secret Monkey Technology (2019).

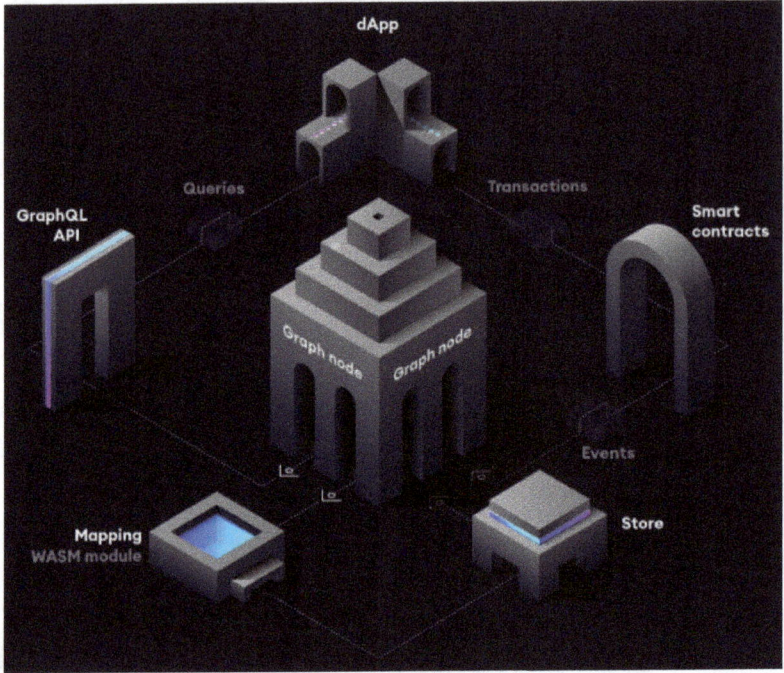

Figure 6.　The Graph's "Roadmap".
Source: Graph Foundation (n.d.).

6.3.　*Data collection and analysis*

Many sites are trying to become the on-chain "Bloomberg" due to easy access to on-chain data. These sites include:

- Glassnode, which provides comprehensive on-chain data for investment and trading decisions in cryptocurrency markets.
- Tokenview, which provides graphical metrics on public chains, NFTs, DeFi, and stablecoins.
- Nansen, which analyzes and tags on-chain activities of Ethereum addresses.
- Dune Analytics, which is community-driven and provides a large number of data analytic dashboards.

For more details, see PANews (2021).

The six sections in this chapter cover the directions of the current mainstream blockchain extension. As the scale of Web3 continues to expand, blockchain and Web3 will need more extensions. More segments would produce economic effects, and it may be prudent to strategize and develop ahead of time.

References

Adams, A. & Ibert, M. (2022). Runs on algorithmic stablecoins: Evidence from Iron, Titan, and Steel. https://www.federalreserve.gov/econres/notes/feds-notes/runs-on-algorithmic-stablecoins-evidence-from-iron-titan-and-steel-20220602.html.

alirezaananas. (2020, November 24). Stablecoin supply. Dune.Com. https://dune.com/alirezaananas/stable-coin.

Arweave — A community-driven ecosystem (n.d.). Retrieved February 15, 2023, from https://www.arweave.org/.

Band protocol — Cross-Chain data oracle (n.d.). Band protocol — Cross-Chain data oracle. Retrieved February 15, 2023, from https://bandprotocol.com.

Benet, J. (2014). IPFS — Content addressed, versioned, P2P file system (arXiv:1407.3561). arXiv. https://doi.org/10.48550/arXiv.1407.3561.

Breidenbach, L., Cachin, C., & Chan, B. (2021). Chainlink *2.0:* Next steps in the evolution of decentralized oracle networks. Chainlink. https://research.chain.link/whitepaper-v2.pdf.

BrightID main LLC. (n.d.). Brightid-meet. Retrieved February 12, 2023, from https://meet.brightid.org/#/.

Browne, R. (June 24, 2022). $100 million worth of crypto has been stolen in another major hack. *CNBC.* https://www.cnbc.com/2022/06/24/hackers-steal-100-million-in-crypto-from-harmonys-horizon-bridge.html.

Buterin, V. (January 26, 2022). Soulbound. https://vitalik.ca/general/2022/01/26/soulbound.html.

Chainalysis. (February 3, 2022). Wormhole hack: Lessons from the wormhole exploit. Chainalysis. https://blog.chainalysis.com/reports/wormhole-hack-february-2022/.

Chainlink Foundation. (2023). Cross-Chain Interoperability Protocol (CCIP). Chainlink. https://chain.link/cross-chain.

Chainlink. (n.d.). What is the blockchain oracle problem? L Chainlink. Retrieved from https://chain.link/education-hub/oracle-problem.

Chainlink. (n.d.). Introduction to Chainlink VRF. Chainlink documentation. Retrieved February 15, 2023, from https://docs.chain.link/vrf/v2/introduction/.

Chainlink. (n.d.). The industry-standard Web3 services platform. Retrieved February 15, 2023, from https://chain.link/.

Circle Internet Financial Limited. (2023). Circle |USDC payments, treasury management, & developer tools. https://www.circle.com/en/.

Circle Internet Financial LLC. (2022, June 16). Circle announces a fully-reserved, Euro-backed Stablecoin, Euro Coin (EUROC). https://www.prnewswire.com/news-releases/circle-announces-a-fully-reserved-euro-backed-stablecoin-euro-coin-euroc-301569124.html.

Dale, B. (December 13, 2018). Basis stablecoin confirms shutdown, blaming 'Regulatory Constraints'. https://www.coindesk.com/markets/2018/12/13/basis-stablecoin-confirms-shutdown-blaming-regulatory-constraints/.

Damgård, I. B. (1990). A design principle for hash functions. In Brassard, G. (Ed.), *Advances in Cryptology — CRYPTO' 89 Proceedings,* pp. 416–427, Springer, New York, NY. https://doi.org/10.1007/0-387-34805-0_39.

Di Nicola, V., Longo, R., Mazzone, F., & Russo, G. (2020). Resilient custody of crypto-assets, and threshold multisignatures. *Mathematics,* 8(10), Article 10. https://doi.org/10.3390/math8101773.

empty set. (n.d.). EmptySet DAO. Retrieved February 15, 2023, from https://www.emptyset.finance/.

Filecoin. (n.d.). A decentralized storage network for humanity's most important information. Filecoin. Retrieved February 12, 2023, from https://filecoin.io/.

Goldston, J., Chaffer, T. J., Osowska, J., & Goins II, C. von (2023). Digital inheritance in Web3: A case study of soulbound tokens and the social recovery pallet within the Polkadot and Kusama Ecosystems (arXiv:2301.11074). arXiv. https://doi.org/10.48550/arXiv.2301.11074.

Graph Foundation. (n.d.). About the graph. The graph docs. Retrieved February 15, 2023, from https://thegraph.com/docs/.

Gu, S. (May 1, 2022). How does oracle make Web3 a better place? Medium. https://medium.com/@sally_gu/how-does-oracle-make-web3-a-better-place-32fa93d4c21e.

Huang, S., Hussain, R., Carletti, S., Liu, D., & Liu, H. (n.d.). BNB smart chain. https://github.com/bnb-chain/whitepaper/blob/master/WHITEPAPER.md.

HTTP Documentation. (n.d.). IETF HTTP working group. February 12, 2023, https://httpwg.org/specs/.

Juels, A. (2020, September 11). Fair sequencing services: Enabling a provably fair DeFi ecosystem. Chainlink Blog. https://blog.chain.link/chainlink-fair-sequencing-services-enabling-a-provably-fair-defi-ecosystem/.

Kanani, J., Nailwal, S., & Arjun, A. (n.d.). GitHub — Maticnetwork/Whitepaper: matic whitepaper. February 12, 2023, https://github.com/maticnetwork/whitepaper/.

Kannengießer, N., Pfister, M., Greulich, M., Lins, S., & Sunyaev, A. (2020). Bridges Between Islands: Cross-Chain technology for distributed ledger technology. Hawaii international conference on system sciences 2020 (HICSS-53). https://aisel.aisnet.org/hicss-53/os/blockchain/4.

Lee, D. K. C., Yan, L., & Wang, Y. (2021). A global perspective on central bank digital currency. *China Economic Journal,* 14(1), 52–66. https://doi.org/10.1080/17538963.2020.1870279.

NEAR Foundation. (n.d.). NEAR | The OS for an open web. Retrieved February 15, 2023, from https://near.org/blog/statement-in-full-near-foundation-to-fund-usn-protection-programme/.

NEST Protocol. (n.d.). The most important infrastructure after ETH. Retrieved February 12, 2023, from https://www.nestprotocol.org/.

Ou, W., Huang, S., Zheng, J., Zhang, Q., Zeng, G., & Han, W. (2022). An overview on cross-chain: Mechanism, platforms, challenges and advances. *Computer Networks*, 218, 109378. https://doi.org/10.1016/j.comnet.2022.109378.

PANews. (2021, December 17). Summary of commonly used blockchain data tool websites. *PANews*. https://www.panewslab.com/zh/articledetails/1639 654376071749.html.

Pillai, B., Biswas, K., Hóu, Z., & Muthukkumarasamy, V. (2022). Cross-blockchain technology: Integration framework and security assumptions. *IEEE Access*, 10, 41239–41259. https://doi.org/10.1109/ACCESS.2022.316 7172.

Sandor, K. & Genc, E. (2022, June 1). The fall of Terra: A timeline of the meteoric rise and crash of UST and LUNA. https://www.coindesk.com/learn/the-fall-of-terra-a-timeline-of-the-meteoric-rise-and-crash-of-ust-and-luna/.

Secret Monkey Technology. (2019, May 29). What is blockchain browser. https://www.zhihu.com/question/293505338.

Storj Labs Inc. (n.d.). Storj — Make the world your data center. Retrieved February 15, 2023, from https://www.storj.io/.

Tether Operations Limited. (2013). Tether. https://tether.to/en/.

The Monero Project. (n.d.). Getmonero.Org, The Monero Project. Retrieved February 12, 2023, from https://www.getmonero.org//index.html.

Thurman, A. (2022, March 29). Axie infinity's ronin network suffers $625M exploit. https://www.coindesk.com/tech/2022/03/29/axie-infinitys-ronin-network-suffers-625m-exploit/.

Witnet. (n.d.) The decentralized oracle network. Retrieved February 15, 2023, from https://witnet.io/.

XY FINANCE. (2023). XY FINANCE — Full cross-chain aggregator enabling smart routing. https://xy.finance/.

Zahed Benisi, N., Aminian, M., & Javadi, B. (2020). Blockchain-based decentralized storage networks: A survey. *Journal of Network and Computer Applications*, 162, 102656. https://doi.org/10.1016/j.jnca.2020.102656.

Chapter 4

Decentralized Finance

Jincheng Zheng

Decentralized finance (DeFi) is an open global financial system in the Web 3.0 era. It enables users to manage financial assets without tedious procedures and control of third parties. Users can borrow, swap, and generate interest through DeFi to improve capital efficiency. DeFi is the first large-scale application of blockchain.

1. The Difference between DeFi and CeFi

Before the emergence of DeFi,[1] users mainly traded and borrowed on centralized exchanges, where the funds were in custody. In addition, centralized exchanges determined the tokens to be listed, the transaction fees, and the rules users should follow.

According to data in May from The Block Research (2022), Binance remains the dominant spot exchange among all U.S. regulatory-approved onshore exchanges, with 64.1% of the market share. Figure 1 shows FTX surpassed Coinbase for the first time with 10.8% of the market share.

Figure 2 shows Binance, OKX, and Huobi have long been the top three exchanges in market share in the global cryptocurrency trading market. Binance's market share has grown from 32.67% in May 2018 to 76.64% in August 2022. OKX and Huobi's market

---

[1] What is DeFi 2.0 and Why Does it Matter? https://academy.binance.com/en/articles/what-is-defi-2-0-and-why-does-it-matter.

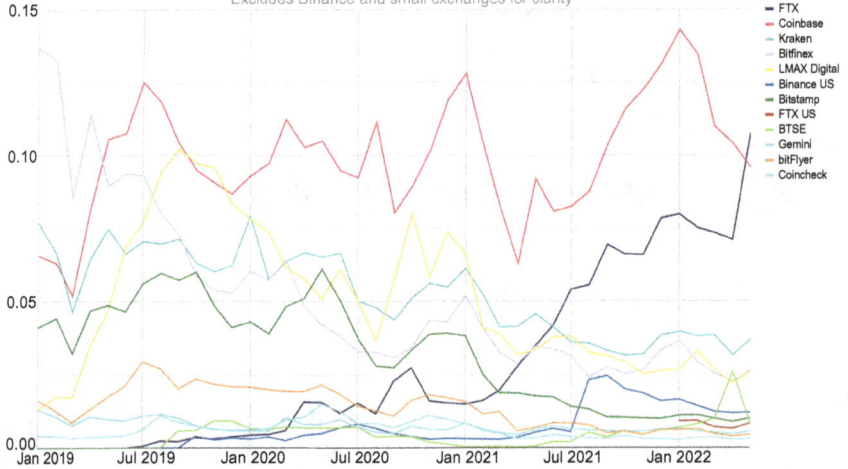

Figure 1.   Market share of legitimate volume spot exchanges.
Source: The Block (2022). Retrieved from https://www.theblock.co/linked/
149654/ftx-surpassed-coinbase-as-second-biggest-centralized-crypto-exchange-
in-may.

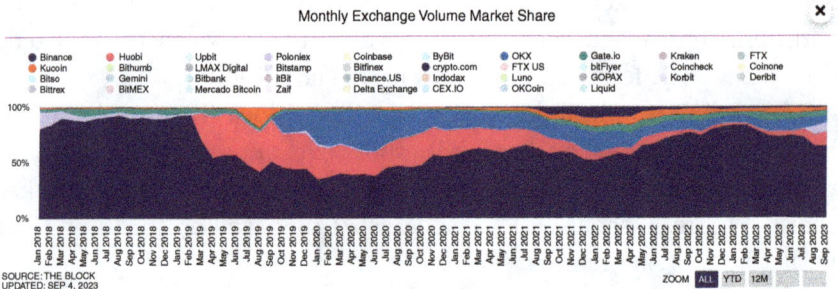

Figure 2.   Crypto-only exchange market share.
Source: *The Block* (2022). Retrieved from https://www.theblock.co/data/crypto-
markets/spot/crypto-only-exchange-market-share.

	DeFi	CeFi
Fund	Self-Custody	Custody
Business	Lending,Payment,Trade,Derivative	Lending,Payment,Trade,Fiat
Customer Information	Permisionless	Need KYC
Risk	Users are responsible for their own fund. Hacking attacks happen often.	CeFi are responsible for clents' fund. Users may suffer loss if CeFi goes bankrupt
Transparency	Fully transparent	Non-transparent
Customer Service	No available	Available

Figure 3.   DeFi vs CeFi.

shares have declined, accounting for 10.91% and 4.89%, respectively. Crypto.com rose in 2021, holding 10% of the market at its best, and then decreased significantly in 2022. Kucoin and Gate.io had 3.8% and 3.5% of the market in August 2022, respectively.

After understanding the current CeFi landscape, let us look at the differences between DeFi and CeFi in Figure 3.

CeFi offers unique fiat services. More DeFi wallets are partnering with payment companies to provide fiat deposit channels, but there is no withdrawal channel. CeFi uses a login model that the general public is more familiar with. Therefore, CeFi is the first stop for newcomers to the crypto world.

However, CeFi requires KYC. There are severe consequences in the event of CeFi data leakage. In addition, CeFi may sometimes deny users' requests to withdraw funds. Furthermore, the financial status of centralized finance is not transparent. Some CeFi such as Celsius,[2] AEX,[3] and Hoo delayed withdrawals in June 2022.

The pain points have prompted users to move digital assets from small exchanges to personal on-chain wallets. Figure 4 shows that the spot transaction volume of decentralized exchanges to the centralized

---

[2]Celsius Network (2022). https://twitter.com/CelsiusNetwork/status/153616901 0877739009.
[3]AEX (2022). https://www.aex.com/announcement/en/blog/2022/06/16/12124. html/?aex-lang=en-GB.

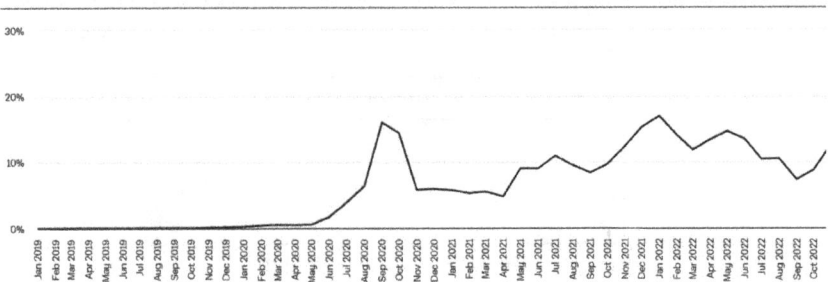

Figure 4.   DEX to CEX spot trade volume.
Source: *The Block* (2022). Retrieved from https://www.theblock.co/data/decen
tralized-finance/dex-non-custodial/dex-to-cex-spot-trade-volume.

exchanges is zig-zagging up, which shows that the blockchain spirit
of trustless and decentralization is beginning to take root.

According to the Paradigm (2022), the liquidity of Uniswap
V3 is twice that of Binance and Coinbase, while the liquidity of
ETH/BTC is three times compared to Binance and 4.5 times that
of Coinbase. These facts show that the automated market maker
model of decentralized exchanges can provide a better service than
centralized order book trading. As on-chain liquidity increases and
blockchain performance improves, there is reason to believe that
permissionless DeFi will produce more financially efficient products.

DeFi and CeFi are competing and pushing each other. CeFi brings
new users into the crypto world, and DeFi keeps people with excellent
products that have the blockchain philosophy. With good CeFi and
DeFi, the crypto market will continue to grow.

2.   DeFi's Development

Born in January 2009, Bitcoin was the first DeFi application, open to
all, giving users the first absolute freedom to control their property
and circulate it worldwide. Bitcoin is a hedge against inflation caused
by uncontrolled money printing with a fixed and scarce supply of
21 million units. While traditional financial markets can be shut
down at some point, Bitcoin's distributed ledger can continue to

operate at all times. Bitcoin's trustless, decentralized open nature is the cornerstone of DeFi.

In 2013, Ethereum brought smart contracts to the blockchain, making cryptocurrencies programmable. Smart contracts perform operations according to pre-defined rules. Since then, the roles of crypto assets are not limited to storing and transferring value.

In November 2017, MakerDAO, which runs on Ethereum,[4] launched Dai. It is an over-collateralized stablecoin that allows users to stake ETH at Maker to mint Dai, which is pegged in USD. It is the first time users receive financing through digital currency.

In September 2018, Compound, the first decentralized lending protocol, went live.

In November 2018, Uniswap, currently, the most prominent decentralized exchange in terms of trading volume, went live with the creative introduction of an automatic market maker mechanism.

In April 2019, TokenSets, a decentralized asset management protocol, went live. It enables users to participate in various investment strategies or become the initiator of a fund.

In July 2019, Synthetix, a decentralized synthetic asset protocol, introduced the concept of "liquidity mining" for the first time.

In June 2020, Compound introduced COMP governance token to reward users in a process known as liquidity mining, sparking the "DeFi Summer 2020".

In January 2020, Curve launched the decentralized exchange, which is the liquidity hub for stablecoins and synthetic assets.

In July 2020, Yearn, the decentralized yield aggregator, pioneered a fair launch, offering token sales to whoever wanted to participate.

Between June and August 2020, the total market capitalization of the DeFi protocol peaked 12 times.

In August 2020, SushiSwap (2022) forked Uniswap, conducting a "vampire attack" by introducing SUSHI tokens to attract liquidity from Uniswap.

---

[4]Ethereum.org. Decentralized finance (DeFi) (2022). https://ethereum.org/en/defi/#main-content.

In September 2020, Uniswap issued "airdrops"[5] to address participating in the protocol, an innovative way of token distribution and incentives to users.

In April 2021, Centrifuge[6] and MakerDAO partnered to provide an 181,000 DAI loan against a house, the first blockchain-based mortgage for a real-world asset.

In May 2021, Uniswap launched V3,[7] which allows liquidity providers to concentrate liquidity and select fee tiers, significantly improving capital utilization.

In November 2021, Aave[8] launched V3, which allows for cross-chain lending, high utilization of the same type of assets and a position-by-position model for risk segregation.

3. Status of DeFi

3.1. *DeFi users surpass 4.8 million*

DeFi users have multiplied over the past few years, demonstrating the demand for alternative financial services. According to Figure 5 from Dune, during the bull market from Q2 2021 to Q2 2022, the number of DeFi users based on wallet addresses grew by 11.2 times. The number of DeFi users grew 6.45 times between Q2 2018 and Q2 2019, the slowest growth period before 2022.

The number of new addresses increased by 1.96 million from Q2 2021 to Q2 2022, bringing the number of users to 4.82 million. But the number of DeFi users increased by only 60% year-on-year, a significant slowdown. In particular, the number of new DeFi users in the first half of 2022 was 630,000, half of the 1.33 million new DeFi users added in the second half of 2021. As a DeFi user may have

---

[5]Uniswap (2020). Introducing UNI, https://uniswap.org/blog/uni.

[6]Lea, S. (2021). DeFi 2.0 — First Real World Loan is Financed on Maker, https://medium.com/centrifuge/defi-2-0-first-real-world-loan-is-financed-on-maker-fbe24675428f.

[7]Hayden, A., Noah, Z., Moody, S., River, K., & Dan, R. (2021). Uniswap, Uniswap v3 Core, https://uniswap.org/whitepaper-v3.pdf.

[8]Emilio (2021). Introducing AAVE V3, https://governance.aave.com/t/intro ducing-aave-v3/6035.

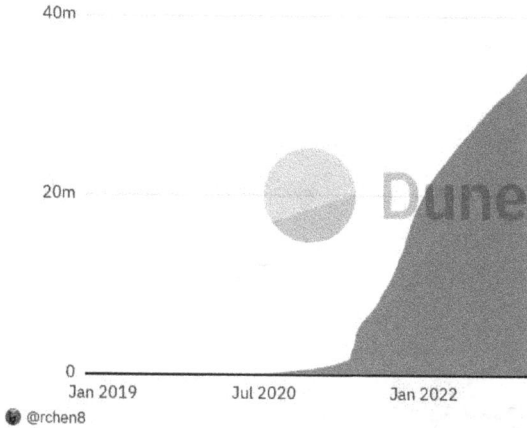

Figure 5. Total DeFi users over time.
Source: *The Block* (2022). Retrieved from https://dune.com/rchen8/defi-users-over-time.

multiple wallet addresses, the actual number of DeFi users is lower than the number based on wallet addresses.

3.2. *The growth speed of DeFi users is slowing*

From Figure 6, the DeFi protocol snowballs between 2020 and 2021 but slows down in 2022. In terms of the number of users, Uniswap is leading the way, with cumulative transaction volumes exceeding 1 trillion on May 22. 1 inch, the decentralized aggregator, ranked second among DeFi protocols. The compound has the most users among lending protocols.

3.3. *TVL in deep correction after rapid growth*

According to data from DeFiLlama, DeFi's total value locked (TVL) exploded 250 times from $1.1 billion on June 1, 2020, to a peak of $184.75 billion on December 1, 2021. An oscillation followed this for five months. The depeg of the algorithmic stablecoin UST on May 8 triggered panic and massive selloffs, causing the TVL of the entire DeFi market to fall off a cliff from $139.31 billion on that day to $84.67 billion on May 14. Some Terra-based protocols were rebuilt in the subsequent Terra 2.0 and partially migrated to other public chains; others were suspended or failed. Prominent hedge fund Three

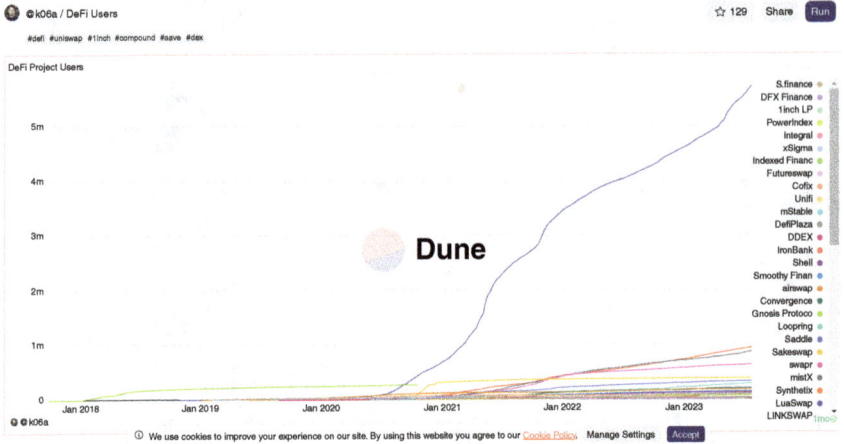

Figure 6.   DeFi Project Users.
Source: *The Block* (2022). Retrieved from https://dune.com/k06a/DeFi-Project-Users.

Arrows Capital also lost much money on Luna and UST. Three Arrows Capital went bankrupt in June, sparking another panic. The sharp drop in the price of the currency triggered a massive liquidation that saw TVL fall rapidly from $77.51 billion on June 11 to $53.29 billion on June 19. Since then, the market has rebounded, and TVL has gradually increased. As of August 10, 2022, DeFi TVL was down 64% from its peak and 37.7% compared to last year (see Figure 7).

3.4.   *Only superpower and multi-great power*

At the beginning of 2020, Ethereum's fees were still friendly to most DeFi users. With the fast growth of the Ethereum ecosystem during the "Summer of DeFi", TVL and new users are multiplying. However, Ethereum's limited TPS cannot meet the demand of users, which leads to high costs and discourages users with less money. These users are looking for a cheaper and faster alternative to Ethereum. According to Figure 8, we can visually see that Ethereum's TVL continued to fall from a peak of 97% to roughly 55% of the overall market.

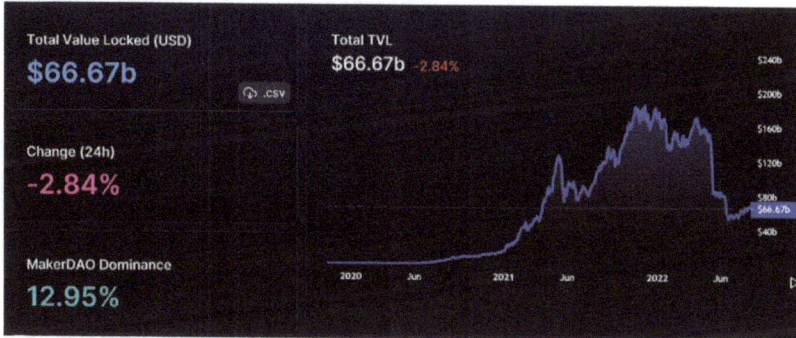

Figure 7. Change in total value locked.
Source: *DeFiLlama*. Retrieved from https://defillama.com/.

Figure 8. Total value locked all chains.
Source: *DeFiLlama*. Retrieved from https://defillama.com/chains.

3.5. *Market chill affects DeFi revenue*

As seen in Figure 9, DeFi protocol revenues are closely related to the market and positively correlate with prices. In June 2022, the major DeFi protocols had total revenues of $80 million. Uniswap received $60.77 million, making it the market leader. The number of staff employed by DeFi protocols is much fewer than in traditional finance for the same volume of business, and the cost of compliance is low so far.

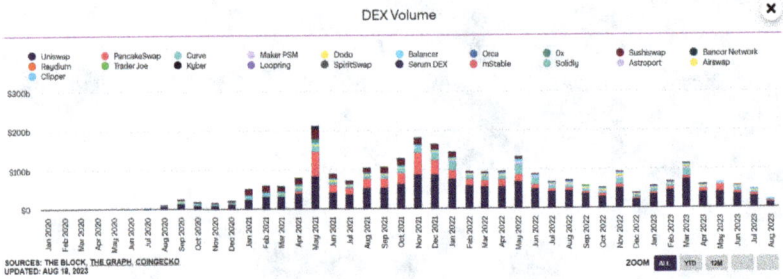

Figure 9. Monthly DeFi revenue.
Source: *DeFiLlama*. Retrieved from https://www.theblock.co/data/decentralized-finance/protocol-revenue/defi-protocols-revenues-monthly.

Figure 10. Price charts for DPI (blue line), DPI/BTC (yellow line), and DPI/ETH (red line).
Source: CoinGeck (2021). Retrieved from https://www.coingecko.com/en/coins/defi-pulse-index.

3.6. *DPI underperforming BTC, ETH*

From Figure 10, we can see the DPI Index, made up of primary blue chip tokens of DeFi, continues to underperform ETH from March 2021 onwards and BTC from May 2021 forward.

The token underperformed from July to November 2021 when revenues were good. In 2022, when revenues declined, the token price trended downwards in accordance with the market. Some investors have said that "DeFi is dead". Prices fluctuate around value. Let's look at the fundamentals of DeFi's various markets and see if the current price is a good bargain.

4. DeFi's Performance Across the Markets

4.1. *Decentralized lending protocols*

The largest segment of the traditional finance industry is commercial banking. Decentralized lending protocols, considered "commercial banking on-chain", are one of the early markets in DeFi to boost. Decentralized lending protocols offer fixed and variable interest rates. They allow users to deposit assets permissionless to earn interest or leverage positions. The smart contract eliminates the middleman, increasing capital efficiency and reducing costs. SMEs with difficulty borrowing money in traditional business can obtain loans by staking existing assets or future cash flow.

Anchor, the top TVL in the lending market with nearly 20% annual percentage yield for a long time, fell from grace after the UST depeg in May. According to Figure 11, AAVE is now the largest "on-chain commercial bank" with twice the total TVL of the second place JustLend. AAVE has deployed on multiple blockchains. Four of the top 10 protocols are from the Ethereum ecosystem.

Most lending protocols nowadays use the over-collateralization model. In times of market decline, borrowers who fail to increase collateral or repay loans in time suffer liquidations. Because of its open and transparent nature, highly leveraged whales are targets for short sellers.

Much of the market decline in June 2022 was caused by whales selling off mortgages to pay off loans to lower the liquidation threshold. But the significant sell-off of assets led to market panic. Falling collateral prices again led to unhealthy collateral ratios, causing the need for further deleveraging. As such, privacy-decentralized lending protocols are the next niche market. In addition, the collapse of the Three Arrows Capital hedge fund reminds practitioners should

Name		Category	Chains ⓘ	TVL	Mcap/TVL
> 1 AAVE			+6	$7.44b	
2 JustLend (JST)		Lending		$3.41b	0.08614
3 Compound (COM...		Lending		$3.07b	0.14405
4 Venus (XVS)		Lending		$740.35m	0.11946
5 Sushiswap Kashi ...		Lending	+3	$342.55m	
6 Solend (SLND)		Lending		$297.39m	0.05266
7 Benqi Lending (QI)		Lending		$287.77m	0.1599
8 Tectonic (TONIC)		Lending		$265.07m	0.06216
9 Euler (EUL)		Lending		$228.91m	
10 Vires Finance (V...		Lending		$183.89m	0.08507

Figure 11.   Top 10 decentralized lending protocols in August 2022.
Source: *DeFiLlama*. Retrieved from https://defillama.com/protocols/Lending.

be mindful of risk control when engaging in financial innovation.
Decentralized credit lending protocols allow loans to white-listed
entities without needing collateral. Investors should be cautious
about such protocols.

On May 18, 2022, Stani Kulechov, founder of Aave, said that
Aave is internally testing the use of NFTs as collateral at the
Permissionless conference. At the same time, we have seen several
funding news in the market for NFT collateralized lending. NFT is
an easy way to attract new users. Using illiquid NFT as collateral
will help to ease the downward pressure on floor prices and improve
capital efficiency. NFT collateralized lending is a business worth
expanding.

The current decentralized and central lending platforms mainly focus on blue-chip tokens, with long-tail assets still underappreciated. Altcoins holders also demand to lend, and advanced players can make strategic investments by borrowing altcoins. The trustless, permissionless nature of decentralized lending protocols is fit for long-tail assets. Euler Finance is the leading player in this overlooked area.

4.2. *Decentralized exchange*

Both investors and traders need to swap tokens. Before the rise of DeFi, centralized exchanges facilitated the majority of transactions. However, the actual financial status of centralized exchanges was not transparent. Users do not own the tokens. The user would suffer loss if the centralized exchanges broke down or refused withdrawal.

A series of exchanges went bankrupt from June to August 2022. In addition, due to compliance, reputation and technical reasons, only part of tokens can be listed on major centralized exchanges. Some valuable tokens have to pay a high cost to be listed. Furthermore, centralized exchanges are not in line with the spirit of decentralization.

To address these pain points, people started to think about building decentralized exchanges that would allow users to trade peer-to-peer at any time to protect their cryptocurrencies. Uniswap is a huge success in this field. It invented an automated market maker mechanism that allowed anyone to act as a market maker to provide liquidity to a pool of tokens. More and more users are withdrawing their tokens from centralized exchanges.

According to Figure 12, four of the top 10 decentralized exchanges in TVL are deployed to the Ethereum ecosystem. Uniswap has the highest TVL, and it is currently the decentralized exchange with the highest revenue, most users, and enormous trading volume. Based on data from Crypto Fees on June 23, 2022, Uniswap's[9] fees exceed Ethereum. Curve, in second place of TVL, is a decentralized exchange focused on the stablecoin and synthesis of assets. Numerous

---

[9]Uniswap (2022). Uniswap has acquired genie, https://twitter.com/Uniswap/status/1539306956002820096

Name		Category	Chains ⓘ	TVL	Mcap/TVL
🔖 1	Uniswap (UNI)	Dexes	⬗ ∞ 🌀 OP ◎	$6.56b	0.63753
🔖 2	Curve (CRV)	Dexes	⬗ ∞ 🌐 🔺 +6	$6.29b	0.08829
🔖 3	PancakeSwap (CAKE)	Dexes	⊛	$3.37b	0.21534
🔖 4	Balancer (BAL)	Dexes	⬗ ∞ 🌀	$1.65b	0.15532
› 5	SUN		▽	$1.28b	0.05322
🔖 6	SushiSwap (SUSHI)	Dexes	⬗ 🌀 ∞ 🔺 +11	$860.22m	0.35597
🔖 7	VVS Finance (VVS)	Dexes	Ⓒ	$749.49m	0.24189
🔖 8	DefiChain DEX (DFI)	Dexes	💲	$403.76m	1.37289
🔖 9	Quickswap (QUICK)	Dexes	∞	$373.89m	0.10841
🔖 10	BiSwap (BSW)	Dexes	⊛	$321.62m	0.35323

Figure 12.   Top 10 decentralized exchanges.
Source: *DeFiLlama*. Retrieved from https://defillama.com/protocols/Lending.

emerging stablecoins and wrapped tokens rely on the Curve's liquidity pool to remain pegged. SushiSwap (2022) is by far deployed on most chains. Pancakeswap is a standout on the Binance Smart Chain.

In the past, decentralized exchanges focused on fungible tokens. The decentralized exchange (DEX) aggregator ParaSwap announced on June 20 it is releasing a peer-to-peer NFT trading app. On June 22, Uniswap announced the acquisition of Genie, an NFT aggregator. In the future, we will see more decentralized exchanges expanding their reach into the non-fungible token market.

4.3.   *Decentralized yield aggregators*

Decentralized Yield Aggregators play a vital role in DeFi by leveraging different protocols and strategies to maximize user profits. However, the security of the new protocols has not yet been tested. So they require professionals to look at the code. In this context,

Name			Category	Chains ⑦	TVL	Mcap/TVL
🔖 1	🔵	Yearn Finance (YFI)	Yield Aggregator	+2	$652.22m	0.55054
🔖 2	🔵	Beefy (BIFI)	Yield Aggregator	+15	$340.52m	0.11454
🔖 3	🟡	Badger DAO (BAD...	Yield Aggregator	+4	$113.63m	0.4222
🔖 4	⚫	Idle Finance (IDLE)	Yield Aggregator		$96.35m	0.01514
🔖 5	↯	Yield Yak (YAK)	Yield Aggregator		$76.05m	0.04981
🔖 6	🔵	Autofarm (AUTO)	Yield Aggregator	+14	$66.11m	0.40113
🔖 7	🔵	Flamincome (FLAG)	Yield Aggregator		$60.28m	
🔖 8	⚪	Rari Capital (RGT)	Yield Aggregator		$49.48m	1.29883
🔖 9	🟣	Vesper (VSP)	Yield Aggregator	+2	$43.04m	0.11245
🔖 10	⚡	Spool Protocol (S...	Yield Aggregator		$40.63m	0.09819

Figure 13.   Top 10 decentralized yield aggregators.
Source: *DeFiLlama*. Retrieved from https://defillama.com/protocols/Yield.

decentralized yield aggregators have evolved. Their roles are similar to that of actively managed hedge funds, helping users to find the best investment opportunities. Key strategies include pooling multiple small funds to reduce gas costs, automatic reinvestment, asset rebalancing and taking advantage of cost-effective mining opportunities.

According to Figure 13, Yearn Finance (2022), Beefy Finance, and Badger DAO (Badger, 2021) are the top three yield aggregators. Beefy Finance and AutoFarm have deployed on multi-chains. Yield Yak is the only aggregator in the top 10 protocols that has not expanded into the Ethereum ecosystem.

Yield aggregators are more vulnerable to hacking because they derive high returns from riskier protocols. According to Coingecko, Yearn Finance (2022) and Harvest Finance suffered losses. According to Badgerdao (Badger, 2021), Badgerdao was attacked due to "unauthorized API Key allowed malicious snippet to set user Web3 permission to attacker wallet".

According to data from Defillma, TVL for the top 10 aggregators (in USD) fell sharply in June, generally by 40–50%, with the most significant drop of 75.17%. To exclude the impact of the fall in cryptocurrency prices, the TVL (in ETH) Yearn Finance (2022), for example, was 1.73 million ETH on January 25, 2022, 655,000 ETH on May 27, 469,000 ETH on June 27, and 339,000 ETH on August 13, showing a steep decline in TVL trend.

4.4.   *Decentralized derivative protocols*

A mature trading market has a much larger volume of derivatives trading than the spot. However, crypto is still a nascent market, so derivatives trading mainly happens on centralized platforms such as Binance, FTX, Deribit and others. There is still much space for decentralized derivatives to develop. This market consists of decentralized perpetual contracts, options, and synthetic assets.

According to Figure 14, Ethereum continues to dominate the derivatives market, with eight of the top 10 protocols deployed on Ethereum or its Layer 2 ecosystem; BNB Chain's derivatives market is growing strongly, attracting four of the top 10 TVL protocols.

dYdX is the flagship Ethereum-based perpetual contract protocol. It partnered with Starkware in the first quarter of 2021 to create a Layer 2 trading solution for faster and cheaper transactions. According to dYdX (2022), "dYdX V4 will be developed as a standalone blockchain based on the Cosmos SDK and Tendermint Proof-of-stake consensus protocol".

According to Wade (2022), the options protocol Opyn introduced Squeth in January 2022, which tracks the ETH^2 index rather than ETH. Compared to a perpetual contract, Squeeth is a leveraged position in the form of an ERC-20 token, where the long position neither strikes nor expiries. Centralized exchanges do not offer this way of combining options with multipliers and perpetual.

The actual trading volumes in the on-chain options and futures are relatively low. The reason is that the current speed of on-chain trading is not satisfying, and the liquidity is not deep enough. In addition, decentralized options and futures protocols would not be able to provide a differentiated service compared to centralized

Name			Category	Chains ⓘ	TVL	Mcap/TVL
🔖 1	✖	Synthetix (SNX)	Derivatives		$679.39m	1.4139
🔖 2		Keep3r Network (KP3R)	Derivatives		$577.74m	0.10923
🔖 3		dYdX (DYDX)	Derivatives		$515.31m	0.54883
🔖 4	▲	GMX (GMX)	Derivatives		$354.65m	0.97276
🔖 5	◆	Perpetual Protocol (PE...	Derivatives		$22.8m	3.38501
🔖 6		ApolloX (APX)	Derivatives		$13.5m	1.47986
🔖 7		Gains Network (GNS)	Derivatives		$11.75m	5.53219
🔖 8		Linear Finance (LINA)	Derivatives		$7,591,063	5.86608
🔖 9		Jarvis Network (JRT)	Derivatives	+2	$7,533,866	0.70298
🔖 10		Deri Protocol (DERI)	Derivatives		$5,601,422	0.64878

Figure 14. Top 10 decentralized derivative protocols.
Source: *DeFiLlama*. Retrieved from https://defillama.com/protocols/Derivatives.

exchanges. An important reason for the development of decentralized exchanges is that they allow anyone to add liquidity pools. Those altcoins that are not listed on a centralized exchange can be traded on the chain. However, long-tail tokens are not suitable for options and futures trading.

4.5. *Decentralized insurance protocols*

In the world of blockchain, "code is law". Hacking and breaches occasionally occur, resulting in enormous losses for investors. A sound insurance system is vital in bringing DeFi to the general public. Decentralized insurance protocols cover the following issues mainly:

Management risk: Private keys are being lost or stolen for protocols that are not fully decentralized.

Technical risk: Smart contracts are vulnerable and subject to hacking.

Impairment losses: Compensation to liquidity providers.

Liquidation of bad debts: In extreme market conditions, there is
a risk that over-collateralized loan protocols will not liquidate assets
on time, resulting in bad debts.

According to Figure 15, Ethereum's massive TVL attracts all the
top 10 insurance protocols, with Armor and Nexus Mutual having a
distinct advantage.

Decentralized insurance is still an inactive market at the moment,
with not enough participants. Retail investors are unwilling to
pay the insurance fee or do not have the time to understand the
mechanisms. The current start-up phase is mainly subsidized by
tokens, failing to form a good balance between supply and demand,
which results in difficulties in pricing insurance. As retail investors
and institutions have experienced or witnessed frequent security
incidents, they will be aware of the importance of this market.

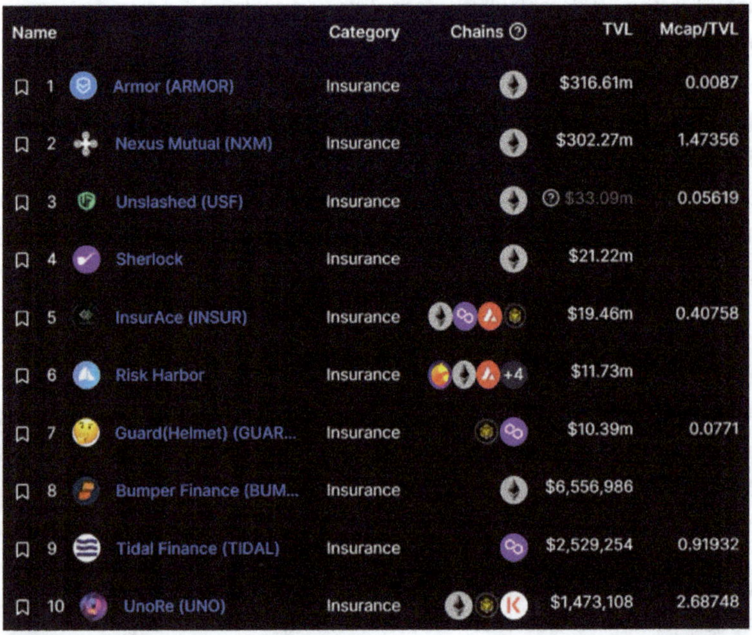

Name			Category	Chains ⓘ	TVL	Mcap/TVL
🔖 1	🛡️	Armor (ARMOR)	Insurance	◈	$316.61m	0.0087
🔖 2	✛	Nexus Mutual (NXM)	Insurance	◈	$302.27m	1.47356
🔖 3	🛡️	Unslashed (USF)	Insurance	◈ ⓘ $33.09m		0.05619
🔖 4	✅	Sherlock	Insurance	◈	$21.22m	
🔖 5	✳️	InsurAce (INSUR)	Insurance	◈ ∞ ⅄ ✳️	$19.46m	0.40758
🔖 6	🔺	Risk Harbor	Insurance	🟠 ◈ ⅄ +4	$11.73m	
🔖 7	🙂	Guard(Helmet) (GUAR...	Insurance	✳️ ∞	$10.39m	0.0771
🔖 8	🅱️	Bumper Finance (BUM...	Insurance	◈	$6,556,986	
🔖 9	🥟	Tidal Finance (TIDAL)	Insurance	∞	$2,529,254	0.91932
🔖 10	🐙	UnoRe (UNO)	Insurance	◈ ✳️ Ⓚ	$1,473,108	2.68748

Figure 15. Top 10 decentralized insurance protocols.
Source: *DeFiLlama*. Retrieved from https://defillama.com/protocols/Insurance.

4.6. *DeFi 2.0 and DeFi 3.0*

DeFi 1.0 uses "liquidity farming" to incentivize users to become market makers. However, short-term incentives create a permanent expense on the balance sheet of the protocol. Users chase high APY mining opportunities without the promise of being loyal, often causing a vicious cycle of "farming and selling".

Therefore, it became the goal of DeFi to develop a better liquidity solution further. Building on DeFi 1.0, the DeFi 2.0 and DeFi 3.0 entered the spotlight.

According to Chase (2021), DeFi 2.0 consists of Liquidity-as-a-Service, Automation-as-a-Service, Enhancement-as-a-Service, and Extenders-as-a-Service.

Liquidity-as-a-Service refers to protocols that buy liquidity directly from the market through a liquidity service provider or rent liquidity from a protocol that offers cheap and high-quality liquidity. Olympus DAO[10] is a protocol to offer liquidity-as-a-service. It incentivizes liquidity providers to sell liquidity to Olympus DAO to get the protocol's token OHM at a discount in the future. This is equivalent to bonding the liquidity. However, Olympus DAO's (3,3) model requires a constant flow of new funds and participants to be loyal holders to maintain high yields. Otherwise, it becomes a situation where the slow runners provide exit liquidity for the fast runners. In a prisoner's dilemma, its token price has fallen 99% from highs, and treasury funds backing OHM have been declining.

Automation-as-a-Service means automating a small part of DeFi. Convex Finance[11] is an excellent project for automation-as-a-service. According to Convex Finance's document, it absorbs users' CRV and Curve LP tokens and helps them to lock, vote, and reinvest automatically, distributing additional CVX tokens to boost returns.

Enhancement-as-a-Service means reusing the fruits of the DeFi 1.0 protocol to provide users with a more optimized solution. One of the fastest-growing protocols is Abracadabra. money, an over-collateralized Stablecoin Protocol. Abracadabra minted the

---

[10]What is Olympus? https://docs.olympusdao.finance/main/basics/readme.
[11]Convex Finance. https://docs.convexfinance.com/convexfinance/.

stablecoin MIM using yield farming tokens such as yvUSDT and yvUSDC as collateral. The liquidation threshold decreases over time as the interest increase the value of collaterals. But the Abracadabra is not decentralized and transparent so far.

Extenders-as-a-Service uses existing DeFi to achieve new functions. For example, users can borrow a synthetic asset form of the underlying collateral, which allows the collateral and the loan to be denominated in the same underlying asset, eliminating the possibility of the collateral being liquidated.

DeFi 2.0 and DeFi 3.0 hit a temporary lull due to the token economics model, contracts issues or personnel management concerns. However, we still see a steady stream of innovative DeFi's emerging on the market, working to solve existing pain points.

5. Current Problems of DeFi

5.1. *Frequent security incidents and substantial financial losses*

In the world of DeFi, the code is the law. However, most users cannot check the code and rely on the protocol for security. If there is a breach of the contract, it can result in huge losses. Once the stolen funds enter the Tornado cash, the chances of recovering them are very low. Cross-chain bridges and yield aggregators are high-incidence areas. In addition, if the team members of protocols lose private keys, it isn't easy to check whether they did it intentionally. In the case of anonymous projects, the likelihood of problems is even more significant, and it is even more challenging to track down responsibility.

Sam (2022) revealed that at least 10 browser plug-in wallets, including MetaMask, had security vulnerabilities in June 2022. Many users had to abandon wallets they had used for a long time. We can see that the infrastructure of DeFi is not yet perfect.

5.2. *Over-innovation and lack of risk awareness*

While traditional financial markets have regulators to control risk, DeFi places more emphasis on financial innovation. The brutal growth of DeFi has resulted in risky projects due to over-innovation.

The (3,3) model, popular in late 2021, has been proven that "making the Ponzi scheme workable" is a false proposition. The collapse of the algorithmic stablecoin UST in May 2022 left a host of institutions and investors out of pocket, causing the dominoes to collapse. Three Arrows Capital's credit loan problems blew up in June 2022, reminding market participants that on-chain credit lending is a minefield to be wary of.

The root of DeFi lies in the fundamentals of finance. The complex token economics often hides risk. Stakeholders should discipline themselves to promote the healthy development of the industry. Investors should not FOMO innovative projects without understanding the mechanics.

5.3. *Liquidity fragmentation in a multi-chain landscape*

Ethereum is the dominant player, but its high costs at peak times are overwhelming for most users. New public blockchains with lower transaction fees and faster speeds have taken up part of the market shares. With the support of major investors, developers and users, ecosystems of some new blockchains have had outstanding achievements in the past year. However, due to technology and competition, most public blockchains are not interoperable, leaving users, assets, data, and DAPPs sealed off within their ecosystems. This is contrary to the spirit of interoperability and scalability of blockchain. The current multi-chain landscape is like a single computer not connected to the internet. There is still much potential to be unlocked.

In this context, the demand for cross-chain is increasing, and various cross-chain solutions are being introduced to the market. However, there is much bad news that happened from cross-chain solutions. Ronin Bridge (2022) announced its validation nodes were compromised in May 2022, resulting in a loss of approximately $600 million. According to Matthew (2022), a hacker compromised at least two out of four private keys of the Horizon Bridge validators, resulting in a roughly $100 million loss. Nomad Bridge (2022) reported it was stolen $190 million on August 1, a few days after announcing to receive investments.

Vitalik[12] said that multi-chain is the future but is pessimistic about cross-chain. He said if the blockchain suffers a 51% attack, the native tokens are unaffected, but the wrapped tokens may not be fully backed. Furthermore, the vast amount of liquidity incentivizes hackers to launch attacks.

5.4. *Inconsistency between token price and protocol value*

The DeFi business acts as a money-printing machine to make much revenue. However, to avoid being classified as securities, most protocols do not pay dividends to holders. Tokens are held with governance rights only. Most retail investors are not interested in governance or do not have the time or energy to follow through, resulting in the protocols being controlled by whales or core teams. So it is difficult for governance rights to support the price of tokens.

The token price has a long-term downward trend. In contrast, Curve's token mechanism captures the value of the protocol better. According to Curve's document (2022), it uses governance tokens to vote on the weight of liquidity incentives. Pools that receive more votes are rewarded with more $Crv. Users are more willing to provide liquidity in pools with more rewards, facilitating the stability of new stablecoin and wrapped token prices. Stakeholders strongly demand to buy $Crv to maintain the token price stable. According to the result of Snapshot in August 2022, Uniswap[13] has reached a consensus to distribute fees to token holders.

Therefore, while avoiding being deemed as securities, the protocols should consider how to make the tokens capture the value of the protocols. Protocols should not ignore the interests of the token holders. The community can be dynamic and sustainable if

---

[12]Vitalik B. Reddit (2022). https://old.reddit.com/r/ethereum/comments/rwojtk/ama_we_are_the_efs_research_team_pt_7_07_january/hrngyk8/.
[13]Leighton (2022). [Consensus Check] "Fee Switch" Pilot, https://gov.uniswap.org/t/consensus-check-fee-switch-pilot/17384.

the protocols and token holders win together. Investors had better not invest blindly.

5.5. *The number of actual users is in doubt*

The retroactive airdrop pioneered by Uniswap gives crypto users a fair way to participate and incentivizes crypto users to test new products. The teams refine the product based on the feedback and eventually receive funding from venture capitalists by showing significant user engagement. Institutional investors find projects that have the potential to grow. Users spend money and time testing new products for free, equivalent to mini-early investors. After receiving funding, teams reward back to the early users. This mechanism is beneficial to all related stakeholders.

However, motivated by the vast wealth effect of airdrop rewards, airdrop hunters have arisen. Airdrop hunters transfer the money to their following wallets after testing protocols. Some programmers have even developed programs to test at hundreds of wallet addresses automatically. As a result, there is no longer an accurate way to assess the actual number of users, and it is more difficult to give a reasonable valuation. After the airdrop rewards were given out, airdrop hunters sold off tokens in large numbers, and the price of the tokens dropped all the way down, hurting investors who had bought tokens from the second market.

In May 2022, Optimism and Hop Protocol announced a crackdown on airdrop hunters and removed their airdrops to curb this phenomenon. In the future, we will see protocols scrutinize the authenticity of their users more closely when giving out airdrops, referencing more on-chain behavior such as whether they have made Gitcoin donations, whether they have voted on governance and so on. Zksync has even publicly stated in discord that it will check the IP addresses of its users.

In addition to the airdrop mechanism, protocols also use subsidies to attract users at early launch. For example, users can earn reward for trading on dydx exchange or listing NFTs at Looksrare. It is worth considering how protocols can retain users with excellent services and reasonable token economics.

6. The Future of DeFi

6.1. Regulation is coming, and traditional financial institutions will develop DeFi businesses

The collapse of the UST in May caused many people to lose much money in a single day. On May 10, 2022, US Treasury Secretary Yellen spoke at a hearing of the Senate Banking Committee about the UST depeg and highlighted the regulations for stablecoins. In June, Three Arrows Capital was serially liquidated and left insolvent. On June 23, 2022, Sopnendu Mohanty, Chief Fintech Officer of the Monetary Authority of Singapore, said MAS "have no tolerance for any market bad behaviour. If somebody has done a bad thing, we are brutal and unrelentingly hard".

The lack of self-regulation in the uncontrolled growth of the industry eventually brought about regulation. As it turns out, even good technology needs rule to guide it. Governments will introduce CBDC to capture the market share of private stablecoins which are not so stable. Some native DeFi protocols may embrace regulation actively. After the regulatory framework becomes clearer, traditional financial institutions will enter DeFi. Using blockchain technology, conventional financial institutions will become more transparent and efficient. At the same time, they will also bring resources from traditional industries to provide "liquidity as a service" for DeFi and to match users with native DeFi protocols.

6.2. Introducing real-world assets

Some DeFi protocols are already being proactively regulated to bring in real-world assets. For example, MakerDAO (2022) announced 6s Capital completed a real estate financing deal worth $7.8 million DAI for Tesla. Compound Labs has launched Compound Treasury, a new product for businesses and financial institutions. The Compound Treasury allows non-crypto companies and financial institutions to swap their dollars for USDCs and receive a fixed interest rate of

4% through a partnership with Fireblocks and Circle. In addition, decentralized insurance to provide insurance services for users' real lives is also the next possible growth track.

DeFi's financial efficiency is much faster than traditional finance. But the DeFi market is still a tiny volume compared to the real world. By combining with real-world assets, DeFi can explore a larger market.

6.3. *Privacy DeFi*

DeFi is public and transparent. Anyone who knows the wallet's address can see the number of assets and the history of past transactions. It is the equivalent of everyone in the world being able to check someone's bank statement. This can, in some cases, pose a threat to an individual's life and property. During the down market in June, large investors suffered a spotting blow because their margin was open to seeing. In addition, with regulation looming, some users will also need privacy about their on-chain assets.

According to Nikhilesh (2022), The Office of Foreign Assets Control (OFAC) added Tornado Cash[14] to its Specially Designated Nationals list, which led to all U.S. persons and entities prohibited from interacting with Tornado Cash. Advocating for absolute freedom will be a niche group's self-importance. Privacy protocols must fulfill regulations' requirements so they can continue to develop.

6.4. *Cross-chain DeFi*

As the significant layer one and layer two ecosystems develop, users are no longer limited to Ethereum, and money flows more frequently. Cross-chain DeFi built on cross-chain protocols is a hot topic for future innovation. For example, a decentralized lending protocol could allow users to borrow money on Polygon using Ethereum assets as collateral. This will enable users to repay faster in times of market decline and reduces the high gas costs in the case of congestion. Cross-chain aggregators, for example, have a better

---

[14]Tornado Cash. Wikipedia. https://en.wikipedia.org/wiki/Tornado_Cash.

chance of finding high-yield farms. However, improved security and speed are prerequisites for the development of cross-chain DeFi.

6.5. DeFi 1.0 becomes infrastructure

Some forks have lost users due to bad debts arising from mechanisms or contract loopholes. Several rounds of extreme conditions have tested blue-chip projects such as Uniswap and Aave, and users have recognized their security. A combination of innovations based on blue-chip projects is the next step for financial innovation. For example, decentralized insurance to insure against impairment losses, using existing yield farming tokens as collateral, and creating a Layer 2 option protocol based on synthetic assets. Those protocols that fork blue-chip items without differentiation will gradually disappear.

6.6. DeFi protocols for Web3 investments

The DeFi protocol is well financed but does not pay dividends. In this fast-growing market, we see more and more protocols setting up venture arms to strengthen their moats. Uniswap Labs (2022) announced the launch of Uniswap Labs Ventures to invest in Web3 infrastructure, development tools and other projects on April 11. The Aave founder said on Twitter they are launching Aave Ventures, where each funded team would receive a ticket to rAAVE.

7. Summary

DeFi is the first large-scale application of blockchain technology, birthed in the last bear market, exploding in the early stages of a bull market starting in 2020 and heading toward a peak in late 2021. With the impact of interest rate hikes, DeFi is stepping into a temporary downturn. Some over-innovative projects are being disproven. However, the foundation of DeFi has been built, with well-established infrastructure and a growing user base. As the underlying blockchain technology continues to advance in the future and DeFi education becomes more widespread, DeFi will take off again with even better products.

References

Akash, T. Centralized finance vs decentralized finance. https://www.leewayhertz. com/defi-vs-cefi/.

Badger. (2021). BadgerDAO exploit technical post mortem. https://badger.com/ technical-post-mortem.

Chase, D. (2021). Beyond DeFi 2.0 — What is driving DeFi's current innovation? https://messari.io/report/beyond-defi-2-0-what-is-driving-defi-s-curr ent-innovation.

Coin Gecko, Lucius, F., & Benjamin, H. (2021). How to DeFi (Advanced). https://www.coingecko.com/account/rewards/how-to-defi-advanced? locale=en.

Curve Finance. (2022). Curve documentation. https://curve.readthedocs.io/_/ downloads/en/latest/pdf/.

David, L. (2015). Handbook of digital currency: Bitcoin, innovation, financial instruments, and big data. https://books.google.com.sg/books?hl=zh-CN &lr=&id=RfWcBAAAQBAJ&oi=fnd&pg=PP1&dq=david+lee+kuo+chue n&ots=2NsKLfvduF&sig=VU9tluaJo8oJIqN0zp11YRtgiAE&redir_esc=y# v=onepage&q=david%20lee%20kuo%20chuen&f=false.

David, L. (2020). *Artificial Intelligence, Data and Blockchain in a Digital Economy.* https://www.worldscientific.com/worldscibooks/10.1142/11787# t=aboutBook.

David, L. & Caroline, L. (2019). Blockchain use cases for inclusive FinTech: Scalability, privacy, and trust distribution. https://www.worldscientific.com/ doi/10.1142/S2705109920500030.

David, L. & Linda, L. (2018). *Inclusive FinTech, Blockchain. Cryptocurrency and ICO.* https://www.worldscientific.com/worldscibooks/10.1142/10949# t=aboutBook.

David, L. & Roy L. (2018). Blockchain — From public to private. https://www. sciencedirect.com/science/article/pii/B9780128122822000073.

David, L., Li, G., & Yu, W. (2018). Cryptocurrency: A new investment opportunity? https://jai.pm-research.com/content/20/3/16/tab-pdf-trialist.

David, L., Joseph, L., Kok Fai, P., & Yu, W. (2022). *Applications and Trends in Fintech I.* https://www.worldscientific.com/worldscibooks/10.1142/12578# t=aboutBook.

dYdX. (2022). dYdX V4 — The dYdX chain. https://dydx.exchange/blog/dydx-chain.

Eli, T. (2022). ParaSwap launches peer-to-peer NFT trading app. https://www. coindesk.com/business/2022/06/20/paraswap-launches-peer-to-peer-nft-tra ding-app/.

Emilio. (2021). Introducing AAVE V3. https://governance.aave.com/t/ introducing-aave-v3/6035.

EU Blockchain. (2022). Decentralised finance (DeFi). https://www.eublockchain forum.eu/reports/decentralised-finance.

Felix F. & Brian, W. (2019). Set: A protocol for baskets of tokenized assets. https:// www.setprotocol.com/pdf/set_protocol_whitepaper.pdf.

Hayden, A., Noah, Z., Moody, S., River, K., & Dan, R. (2021). Uniswap, uniswap v3 core. https://uniswap.org/whitepaper-v3.pdf.

Lea, S. (2021). DeFi 2.0 — First real world loan is financed on maker. https://medium.com/centrifuge/defi-2-0-first-real-world-loan-is-financed-on-maker-fbe24675428f.

Leighton. (2022). [Consensus Check] "Fee Switch" pilot. https://gov.uniswap.org/t/consensus-check-fee-switch-pilot/17384.

MakerDAO. (2017). MakerDAO formal white paper. https://makerdao.com/en/whitepaper/sai/#overview-of-the-dai-stablecoin-system.

Matthew, B. (2022). Harmony's horizon bridge hack. Medium. https://medium.com/harmony-one/harmonys-horizon-bridge-hack-1e8d283b6d66.

Michael, E. (2019). StableSwap — Efficient mechanism for Stablecoin liquidity. https://curve.fi/files/stableswap-paper.pdf.

Nakamoto, S. (2008). Bitcoin: A peer-to-peer electronic cash system. https://bitcoin.org/bitcoin.pdf.

Nikhilesh, D. (2022). Crypto-mixing service Tornado cash blacklisted by US Treasury. https://www.coindesk.com/policy/2022/08/08/crypto-mixing-service-tornado-cash-blacklisted-by-us-treasury/.

Nomad Bridge. (2022). Nomad bridge hack: Root cause analysis. Medium. https://medium.com/nomad-xyz-blog/nomad-bridge-hack-root-cause-analysis-875ad2e5aacd.

Paradigm. (2022). The dominance of Uniswap v3 liquidity. https://www.paradigm.xyz/2022/05/the-dominance-of-uniswap-v3-liquidity.

Robert, L. (2020). Compound governance is live, medium. https://medium.com/compound-finance/compound-governance-decentralized-b18659f811e0

Robert, L. & Geoffrey, H. (2019). Compound: The money market protocol. https://compound.finance/documents/Compound.Whitepaper.pdf.

Ronin. (2022). Community alert: Ronin validators compromised. https://roninblockchain.substack.com/p/community-alert-ronin-validators.

Sam, K. (2022). MetaMask, Phantom and other browser wallets patch security vulnerability. https://www.coindesk.com/tech/2022/06/15/metamask-phantom-and-other-browser-wallets-patch-security-vulnerability/.

SushiSwap. (2022). SushiSwap intro. https://help.sushidocs.com/products/sushiswap-pools.

Swee Won, L. Yu, W., & David, L. (2021). *Blockchain and Smart Contracts: Design Thinking and Programming for Fintech*. https://books.google.com.sg/books?hl=zh-CN&lr=&id=fJgZEAAAQBAJ&oi=fnd&pg=PR7&dq=david+lee+kuo+chuen&ots=d2pA5dxgm4&sig=3QPoFbAZZRbJh_6_JxHfFv zUzv8&redir_esc=y#v=onepage&q=david%20lee%20kuo%20chuen&f=false.

Synthetix. (2022). Lightpaper. https://docs.synthetix.io/litepaper/.

Tan Choon, Y., Paul S., & David (2018). LInsurTech and FinTech: Banking and insurance enablement. https://www.sciencedirect.com/science/article/pii/B9780128104415000117.

The Block. (2022). FTX surpassed Coinbase as second-biggest centralized crypto exchange in May (2022). https://www.theblock.co/linked/149654/ftx-surpassed-coinbase-as-second-biggest-centralized-crypto-exchange-in-may.

Uniswap. (2019). The uniswap V1 protocol. https://hackmd.io/C-Dvw DSfSxuh-Gd4WKE_ig.

Uniswap. (2022a). Introducing UNI. https://uniswap.org/blog/uni.

Uniswap. (2022b). Uniswap has acquired genie. https://twitter.com/Uniswap/status/1539306956002820096.

Uniswap Labs. (2022). Uniswap labs ventures. https://blog.uniswap.org/ventures.

Uniswap Team. (2022). Uniswap labs ventures. https://uniswap.org/blog/ventures.

Vitalik, B. (2014). Ethereum: A next-generation smart contract and decentralized application platform. https://ethereum.org/669c9e2e2027310b6b3cdce6e1c5 2962/Ethereum_Whitepaper_-_Buterin_2014.pdf.

Wade, P. (2022). Squeeth primer: A guide to understanding Opyn's implementation of squeeth. https://medium.com/opyn/squeeth-primer-a-guide-to-understanding-opyns-implementation-of-squeeth-a0f5e8b95684.

Wahid, P. (2022). Singapore vows to be 'Unrelentingly Hard' on bad behavior in digital assets market. https://coingeek.com/singapore-vows-to-be-unrelenti ngly-hard-on-bad-behavior-in-digital-assets-market/.

What is DeFi 2.0 and why does it matter? https://academy.binance.com/en/articles/what-is-defi-2-0-and-why-does-it-matter.

Yearn Finance. (2022). V3 white paper. https://www.allcryptowhitepapers.com/wp-content/uploads/2020/12/YFI3.pdf.

Chapter 5

Non-Fungible Token

Kyle Zhao and Jiancang Guo

Non-Fungible Token (NFT) refers to a non-homogenized crypto token, which is unique and irreproducible and has the ability to represent ownership.

Each NFT is different and is essentially a digital asset notarized on the blockchain. This asset can be a picture, an audio, a video, a string of code, a digital signature, and any other form of a digital asset.

1. Main Categories of NFT

1.1. *Art*

The adoption of NFT in the field of art collectibles is one of the most widespread uses of NFT, especially with the rapid development of computationally generated art and NFT. Artists write codes in programming languages that are processed by computers to produce a series of visual results, which are then combined with NFT techniques and properties to enable a new way of presenting and circulating artworks. In March 2021, digital visual artist Beeple's set of NFT artworks, Everydays: The First 5000 Days, went live at Christie's and was sold for a record $69.34 million. It was the world's first purely digital work to be sold at a traditional auction house (Virtue Market Research, 2022).

1.2. *Profile picture (PFP)*

PFPs are one of the most dominant forms of NFT representation, and these NFTs are widely used as profile pictures on social media such as Twitter, and even as a symbol of status for the NFT holder. The most successful PFP NFTs to date are the Bored Ape Yacht Club (BAYC) series created by Yuga Labs, which attracted NBA star Steve Curry, singer Justin Bieber, Jay Chou, and others to become BAYC holders.

Another popular type of PFP NFT is the CryptoPunks, a pixel art avatar project that predates BAYC as a mainstream PFP project, and each of their NFTs is a very personalized punk image. The number of PFP NFT projects now exceeds hundreds, covering a variety of topics, including the female-driven project World of Women, the Japanese manga character project Azuki, the cartoon owl project Moonbirds, and many more.

1.3. *Music*

Music creators make albums into NFTs and sell them directly to fans, increasing the revenue of the creators. While in the Web2 era, most of the revenue from music creators' works was captured by the platforms, in the Web3 era, creators can directly sell their singles or albums in NFT form on music NFT marketplaces such as Sound.xyz and Royal (Guan *et al.*, 2023). These NFTs usually include a creator's share, and each time they are resold, a percentage of the profit goes to the creator, who continues to receive revenue from the transactions.

1.4. *Game props*

Gamers worldwide spend tens of billions of dollars in-game each year, but the assets players accumulate in-game (gear, experience points, characters, etc.) do not belong to them and can be confiscated by the platform at any time. This is completely changed in the blockchain games with NFTs. NFTs serve as a way for players to own digital property rights in-game. Players will have ownership of in-game items and able to retrieve the value they have created from the platform, and users become part of the game's development.

For example, each of the Axies NFTs in the Axie Infinity game has its own unique attributes and characteristics and can be used for in-game combat, breeding, and collecting. There is also the StepN NFT in Move to Earn project, where players must first get hold of a sneaker NFT in the StepN App app, and then earn token rewards by walking or running in the real world. Each shoe NFT has different attribute characteristics, making the user earn rewards at different rates and with different amounts.

1.5. *Membership*

NFTs may serve as a membership certificate that allows the holder to join a specific association or community (e.g., DAO), or as a qualification for token airdrops. For example, Bored Ape (BAYC) is not only available as an art collection, but its holders can also participate in an exclusive BAYC club, which includes getting first dibs on new NFTs, NFT upgrades, and other privileges. In addition, BAYC holders also have airdrop eligibility for APE tokens issued by Yuga Labs.

1.6. *Redeemable physical products*

Redeemable NFTs allow holders to redeem NFTs into real-world physical assets such as shoes, clothes, paintings, houses, cars, wine, or anything that physically exists. This redemption process is typically one-way, and the NFT is destroyed after redemption.

1.7. *Identity*

A digital tokenized identity based on NFT will make your identity easy to verify, bringing an additional layer of security that makes it difficult to fake or forge an identity. Identity NFT also brings additional possibilities such as tying your digital assets, various proofs or certificates, and personal credit scores to identity NFT and the ability to perform KYC authentication without providing details.

1.8. *Virtual land*

The explosion of the metaverse concept has led to the rise of metaverse platforms such as The Sandbox, Decentraland, and others.

Hundreds of traditional brands have started to move in, and to do so, they first need to "buy land" on which the owner can build freely, opening playgrounds, painting exhibitions, concerts, etc. Most of the activities in the real world can take place in the virtual world, and physical boundaries are eliminated.

1.9. *Financial derivatives*

After the launch of Uniswap V3, LP token with a limited price range became NFT, which opened the door to the NFTization of financial derivatives. Futures, options, swaps, insurance, and other non-standardized contracts in traditional financial markets can theoretically be issued and traded on the chain as NFTs.

2. NFT Marketplace Platforms

2.1. *OpenSea*

OpenSea was co-founded by Devin Finzer and Alex Atallah in 2017 and received a seed round of funding from well-known investor Y Combinator (YC) in 2018 (see Figure 1). Since then, the crypto market started a two-year-longear long bear market and it was only in 2020 that the market warmed up to a new bull market in the crypto and the NFT wave came along with it, and OpenSea became the largest NFT marketplace. OpenSea is now the largest

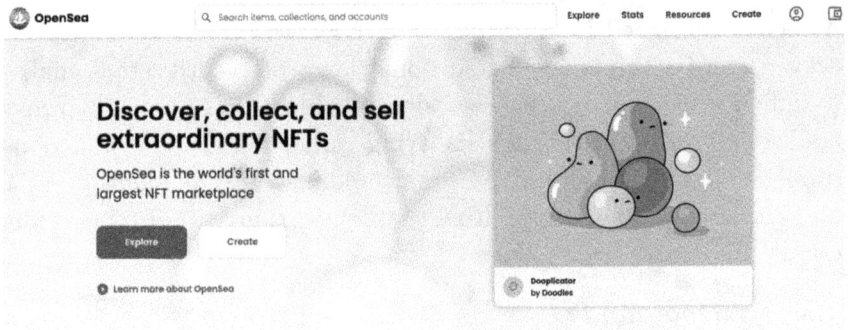

Figure 1.   Front-page of OpenSea website.

NFT marketplace with a total trading volume of over \$30 billion and over 1.8 million traders, while OpenSea's platform valuation has exceeded \$10 billion.[1] As of June 2022, OpenSea has raised \$427.2 million in six rounds of funding from a16z, Paradigm, YC, and others (see Table 1).

Table 1.   OpenSea's six rounds of funding.

Date of funding	Round	Valuation	Fund raised	Investors
January 4, 2022	Series C	\$13.3 billion	\$300 million	Paradigm; Coatue Management
July 20, 2021	Series B	\$1.5 billion	\$100 million	a16z; Coatue; Connect Ventures; Michael Ovitz; Kevin Hartz; Kevin Durant
March 18, 2021	Series A	Unknown	\$23 million	a16z; Naval Ravikant, Mark Cuban, Tim Ferris, Belinda Johnson, Ben Silbermann, Alexis Ohanian; 1confirmation, Pascal capital, Blockchain Capital, Rega Bozmanan, Kevin Hartz and Dylan Field
November 19, 2019	Seed	Unknown	\$2.1 million	gumi Cryptos Capital (gCC); Blockchain Capital; 1confirmation; Animoca Brands; StartX; Dylan Field
May 10, 2018	Seed	Unknown	\$2 million	1confirmation; Blockchain Capital; Coinbase Ventures; Foundation Capital; The Stable Fund; The Chernin Group; Founders Fund
January 4, 2018	Pre-seed	Unknown	\$120k	Y Combinator (YC)

---

[1]OpenSea. OpenSea Whitepaper. https://docs.opensea.io/.

Top NFTs

The top NFTs on OpenSea, ranked by volume, floor price and other statistics.

	All time ∨	⊠ All categories ∨	∞ All chains ∨			

Collection		Volume ‑	24h %	7d %	Floor Price	Owners	Items
1	CryptoPunks	◆ 939,219.64	−37.71%	+319.93%	---	3.5K	10.0K
2	Bored Ape Yacht Club	◆ 618,631.09	−57.29%	−30.11%	◆ 89	6.5K	10.0K
3	Mutant Ape Yacht Club	◆ 421,634.38	+46.78%	−39.55%	◆ 17.55	13.2K	19.4K
4	Otherdeed for Otherside	◆ 304,581.1	+14.74%	−33.71%	◆ 2.43	34.3K	100.0K
5	Art Blocks Curated	◆ 257,970.97	−39.31%	+138.82%	---	11.7K	57.1K
6	Azuki	◆ 251,449.72	−79.89%	+21.19%	◆ 11.9	5.2K	10.0K

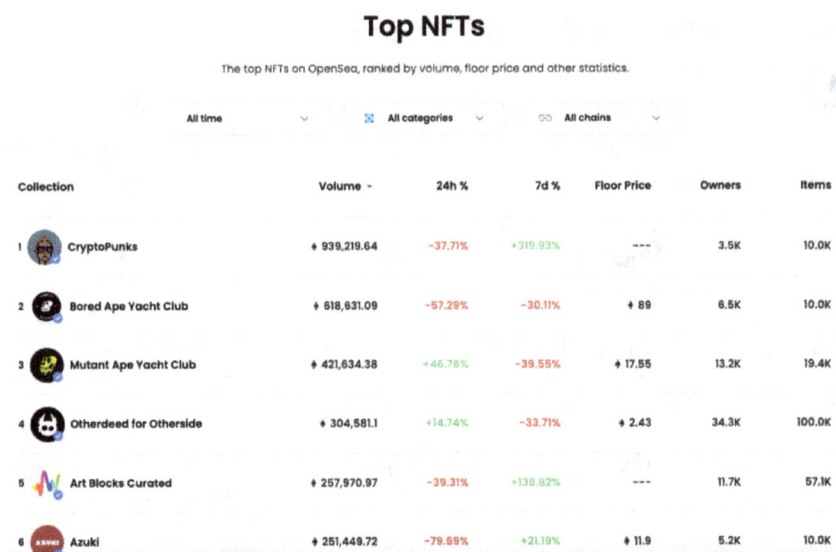

Figure 2.   Ranking of Top NFTs on OpenSea.
Note: Service fee: 2.5% of transaction amount; Platform token: None.
Source: https://opensea.io/.

Initially supporting only NFT projects built on Ethereum, OpenSea has since then added support for Polygon, Solana, and klaytn blockchains as user demand and the NFT ecosystem of other competing blockchains has flourished. OpenSea currently supports connections to 13 different wallets, including MetaMask, Coinbase Wallet, WalletConnect, Phantom, and other major wallets.

OpenSea also offers data analysis services, providing information on the top-ranked NFTs on the platform in terms of their trading volume, reserve price, and other statistics (see Figure 2).

2.2.   *LooksRare*

Founded by two anonymous founders, Zodd and Guts, LooksRare has made it clear from the start that it is a community-first NFT marketplace with a focus on "By NFT people, For NFT people", offering rewards to participating traders, collectors, and creators. Unlike OpenSea, LooksRare issues its own governance token, LOOKS, and offers token airdrops to OpenSea users (see Figure 3). A 2% of user

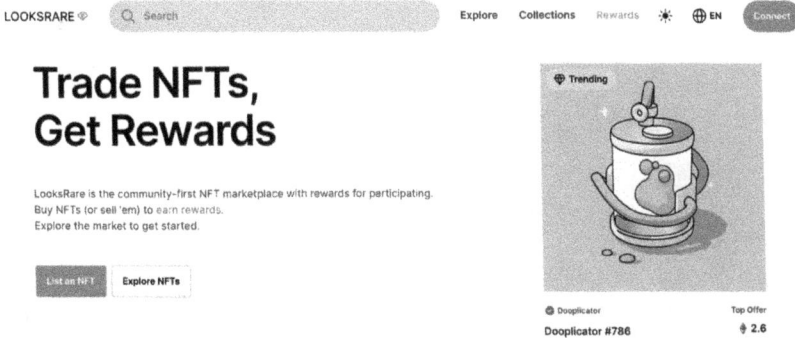

Figure 3.   Front-page of LooksRare website.
Note: Service fee: 2%. Platform token: LOOKS. *Investor*: Assembly partners.
Source: https://looksrare.org/.

transactions are charged by LooksRare as a service fee (0 for private sales), which is distributed in full to users who pledge LOOKS tokens in the form of WETH.

2.3.   *Rarible*

Rarible is an NFT marketplace for creators to distribute and sell their work, including painters, music creators, singers, etc. Rarible is an open source platform that is open to everyone, including veteran artists or fans of related fields (see Figure 4). In addition, even without coding skills, anyone can easily create and shelve their own NFTs.

Royalty model: Rarible allows users to set a creator royalty of 0–50%, meaning that each time they create an NFT and sell it again, they will share in the revenue.

Free casting model: Creators can choose a "free casting" model when casting NFTs, where they can publish their NFTs without paying a casting fee, but the casting fee will be passed on to the buyer, that is, when the buyer buys the NFT, the buyer will pay the cost of the Gas needed to cast the NFT.

In July 2020, Rarible released its own ERC-20 governance token, RARI, which combines digital collectibles with farming revenue and

Figure 4.   Front-page of Rarible website.

Table 2.   Rarible's three rounds of funding.

Date of funding	Round	Valuation	Fund raised	Investors
June 23, 2021	Series A	Unknown	$14.2 million	Venrock Capital, CoinFund and 01 Advisors
February 3, 2021	Seed	Unknown	$1.75 million	1kx; ParaFi Capital, Coinbase Ventures, Bollinger Investment Group, MetaCartel Ventures and CoinFund
September 8, 2020	Pre-seed	Unknown	Unknown	CoinFund

Note: Service fee: Buyer 1% + Seller 1% Service fee. Platform token: RARI.
Source: https://rarible.com/.

mobile mining. Users of the platform are then rewarded with RARI proxy governance coins.

As of June 2022, Rarible has completed three rounds of funding, raising a total of approximately $16 million in funding (see Table 2).

2.4.   *SuperRare*

SuperRare is an exclusive crypto art marketplace that calls itself "Instagram + Christie's" and lists a selection of digital artworks by

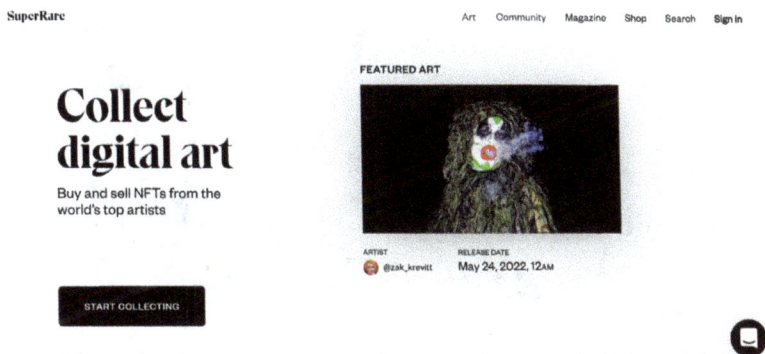

Figure 5.   Front-page of SuperRare website.
Source: https://superrare.com/.

renowned NFT artists and up-and-coming creators (see Figure 5). Unlike other NFTs, which are open and free to enter, SuperRare NFT Marketplace is initially a centralized platform that is rigorously vetted by its core team of incoming artists and has earned a reputation in the NFT platform space as the "premium NFT platform" due to its "quality over quantity".

By June 23, 2022, SuperRare has listed over 35,000 artists' works. In 2021, the platform introduced its RARE governance token — beginning the transition to a DAO led by a community of creatives and collectors.

Funding: On March 30, 2021, SuperRare closed a $9 million Series A round led by Velvet Sea and 1confirmation, with Mark Cuban, Chamath Palihapitiya, and Marc Benioff also participating.

3.   NFT Fragmentation Platform

3.1.   *Fractional*

Fractional is an NFT fragmentation platform that addresses the lack of liquidity in NFT through smart contracts that allow users to split ERC-721-based NFT into multiple ERC-20 tokens, whose ownership is split together to facilitate flow and trading (see Figure 6). Fragmentation means that we can create liquidity on NFT through efficient pricing.

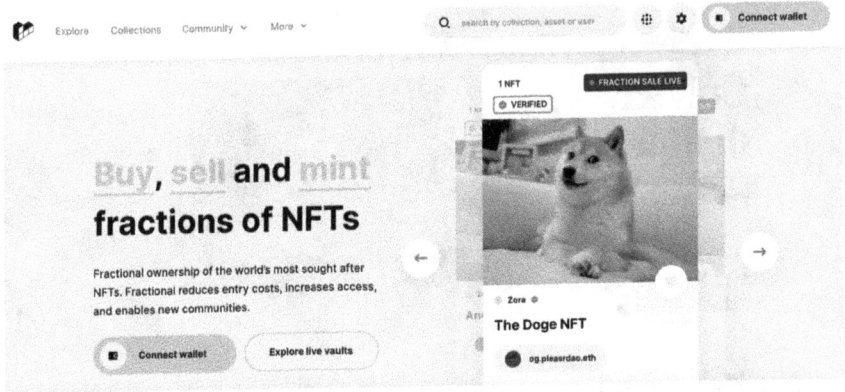

Figure 6.   Front-page of Fractional website.

In the Fractional platform, to fragment an NFT, you need to create a vault (vault) to manage your NFT, set the name and number of tokens to be fragmented, and the floor price.[2]

Once the NFT is fragmented, users who bought the fragmented tokens will have collective ownership of the NFTs locked into the vault and can vote on the reserve price of the NFT assets, which will trigger a third party to initiate an NFT auction once the reserve price is determined. The eventual winner of the participating auction will receive the full NFT, while the assets in the hands of the fragmented token holders will be destroyed and exchanged back to ETH.

In August 2021, Fractional closed a $7.9 million seed round led by Paradigm with participation from Robot Ventures, Divergence Ventures, Flamingo DAO, Variant Fund, and Delphi Ventures.

3.2.   *Unicly*

Unicly is a license-free, community-managed protocol for combining, fragmenting, and trading NFTs, blending DeFi elements with NFTs (see Figure 7). Its functionality combines AMM, NFT auctions, revenue farming and decentralized governance to create an NFT gallery and trading platform, while providing NFT sharding and

---

[2]Explained: Fractional NFTs (F-NFTs) and How They Work. https://learn.bybit.com/nft/what-are-fractional-nfts/.

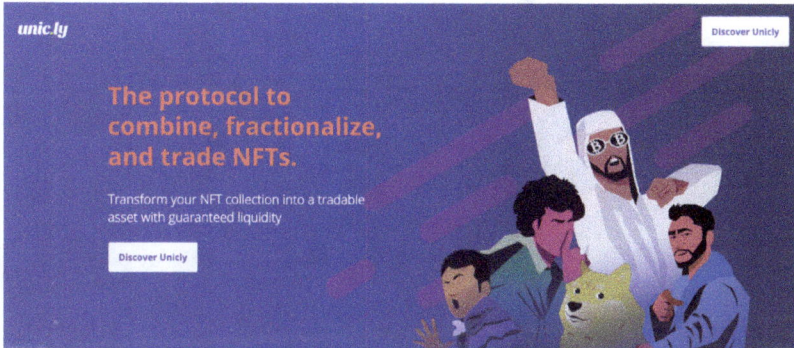

Figure 7.   Front-page of Unicly website.
Source: https://www.unic.ly/.

token trading capabilities. Launched by anonymous creator 0 xLeia, the protocol was built by NFT collectors and DeFi enthusiasts to incentivize NFT liquidity and provide a seamless trading experience for NFT assets.

Similar to Fractional, on the Unicly platform, users can lock their NFTs (any number) into a smart contract and then create their own U-token, customize the token parameters (token name, total amount, etc.), in addition to setting the percentage of tokens needed to unlock the NFTs. This token percentage is what percentage of token votes are required to trigger the activation (bidding) of the NFT for sale.

In addition, users can create liquidity pools for their U-tokens on Unicly's AMM platform UnicSwap.

Funding: On November 28, 2021, Unicly closed a $10 million funding round led by Blockchain Capital, Animoca Brands, and followed by Ascensive Assets, 3 Commas Capital, Morningstar Ventures, and other investors.

4.   NFT Aggregation Platform

4.1.   *Gem*

Gem allows users to purchase NFTs across multiple NFT market-places in a single transaction and can use any ERC-20 token, not just Ethereum (Figure 8). Users can also view analytics such as

Figure 8.   Front-page of Gem website.
Source: https://www.Gem.xyz/.

sales, reserve prices, and rarity-based rankings of NFT collections. The ultimate goal is to consolidate all NFT marketplace pending orders and not have users repeatedly jump across different platforms to compare prices. Currently, it has already aggregated marketplaces including OpenSea, Rarible, LooksRare, X2Y2, and more (Heidorn, 2022).

In terms of data analysis, Gem has partnered with Dune Analytics, a professional data site, to provide professional NFT traders with excellent NFT analysis data.

In terms of NFT rarity, Gem has partnered with Rarity Sniper to display key information such as NFT rarity rankings on the Gem platform.

More importantly, Gem enables cheaper gas costs by offering multiple marketplace aggregations compared to buying on OpenSea, and supports bulk buying/selling of NFTs, which can dramatically improve the transaction experience for buyers looking for a low-price sweep.

On April 25, 2022, OpenSea officially announced the acquisition of Gem, an NFT trading aggregator. OpenSea said that Gem will continue to operate as a standalone brand after the acquisition is completed, and OpenSea will integrate Gem's functionality in the future.

4.2. *Genie*

Genie provides users with a library of NFTs available across markets and enables them to view and trade NFTs on most of these platforms, providing a one-stop NFT trading platform while working to save on trading gas costs and trading time. Genie is often referred to as the first NFT aggregator (Figure 9).

Aggregated display: Genie can display shelf information from major NFT marketplaces, so you can search for more NFT results in fewer tags. These marketplaces include OpenSea, NFTX, Rarible, X2Y2, LooksRare, NFT20, and Larva Labs, among others.

One click sweep: Genie also supports bulk buying and selling, allowing you to quickly sweep markets in a single transaction. Simply go to the collection you want to sweep, click the Sweep button, and then filter the NFTs that meet the sweep criteria you set.

Cost savings: Compared to buying directly from the Marketplace, Genie can save up to 40% on gas fees, depending on the number of NFTs packaged in a single transaction.

Much like Gem, Genie also did not escape the fate of being acquired. On June 21, 2022, Uniswap announced that it had acquired

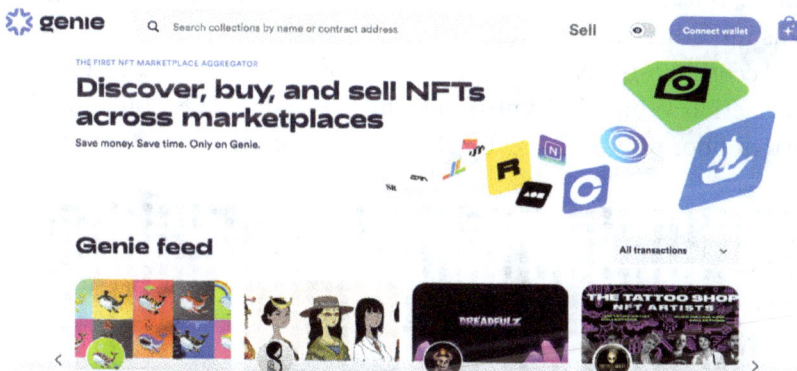

Figure 9. Front-page of Genie website.
Source: https://www.genie.xyz/.

the NFT aggregator Genie. Genie will be integrated with Uniswap at the product level, meaning that Genie will appear in Uniswap's product interface as a feature of NFT aggregation transactions. This feature is expected to go live in the fall. Meanwhile, Uniswap will be airdropping USDC to Genie users in August 2022.

Funding: On March 12, 2022, Genies closed a $150 million funding round led by private equity firm Silver Lake, which valued the company at $1 billion.

In May 2021, Genies closed a $65 million funding round led by Mary Meeker's Bond Capital with participation from Dapper Labs, Polychain, Coinbase Ventures, Hashkey, and others.

5. Major NFT Projects

5.1. *CryptoPunks*

CryptoPunks is one of the first NFT projects on Ethereum. Developed by Larva Labs in June 2017, CryptoPunks are algorithmically generated 24 × 24-pixel art images, as shown in Figure 10. Most are punk-looking boys and girls, but there are some rarer types: apes, zombies, and even weird aliens (see Figure 10). Each punk has its own profile page that shows its attributes and its ownership/pending status. Initially, CryptoPunks are available for free to anyone who owns an Ethereum wallet.

According to NFT.GO data, as of June 24, 2022, the total transaction value of CryptoPunks has exceeded $2.83 billion, with a current floor price of 64.39 ETH (current value of about $73,000). The

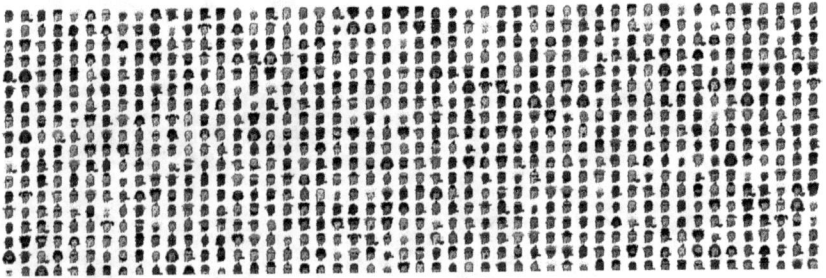

Figure 10.   Front-page of CryptoPunks website.

largest transaction in terms of dollar value occurred on February 12, 2022, when CryptoPunk #5822 was sold for 8,000 ETH (then worth $23.7 million). The largest transaction in ETH terms took place on October 29, 2021, when CryptoPunk #9998 was sold for 124,460 ETH.

CryptoPunks was the inspiration for the modern CryptoArt movement, which has made headlines in mainstream media around the world, and whose creative image, operating model, and success have served as a source of inspiration for many others since.

On March 12, 2022, Bored Ape publisher Yuga Labs announced the acquisition of the CryptoPunks brand, art rights, and other intellectual property from Larva Labs, as well as 423 CryptoPunks.

Current floor price: 64.39 ETH
Total market cap (June 24): $2.06 billion
Total trading volume (in USD): $2.83 billion
All-time high price (in ETH): CryptoPunk #9998 sold for 124,460 ETH on October 29, 2021.

5.2. Bored Ape Yacht Club (BAYC)

Bored Ape Yacht Club (BAYC), or Bored Ape for short, was created by Yuga Labs in 2021 (see Figure 11). It is a collection of 10,000 unique ape NFTs (Lee, 2022). This NFT also doubles as your Yacht Club membership card and grants exclusive member benefits, the first of which is the use of the collaborative graffiti board THE BATHROOM. The community can unlock future areas and benefits by activating a roadmap.

Initially, Bored Ape NFT is available for public sale at 0.08 ETH each. Each Bored Ape is unique and programmatically generated from over 170 possible features, including expressions, headgear, clothing, earrings, hats, fur, backgrounds, and more. When you hold a BAYC NFT, you not only own its collector value, but you also gain membership in a club whose benefits and offerings will grow over time. For example, Otherside, a metaverse game world developed by Yuga Labs, airdrops free metaverse plots NFT Otherdeed to BAYC NFT holders, and Apecoin, developed by Yuga Labs, airdropped

Figure 11.   Front-page of BAYC website.

10,094 APE tokens to each BAYC NFT holder when it launched, which was worth up to $280,000 at the time.

According to NFT.GO data, as of June 24, BAYC NFT total transaction volume reached $2.13 billion, with a current market cap of $1.68 billion and 6456 holders.

Total trading volume: $2.13 billion
Market capitalization (as of June 24): $1.68 billion
Current Floor Price: 87.17 ETH
All-time high price (in ETH): Bored Ape Yacht Club #232 sold for 1080 ETH on January 31, 2022

5.3. *Otherside*

Otherside is a metaverse game that aggregates various NFT projects and is developed by Yuga Labs, the parent company of Board Ape, in collaboration with Animoca Brands, a publisher of NFT games (Figure 12). On April 30, 2022, Yuga Labs announced that it would launch Otherdeed, a new NFT designed specifically for the game. The NFT will serve as a credential for virtual land in the Otherside

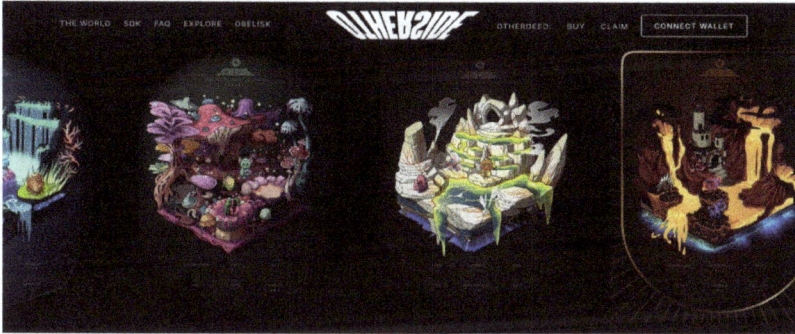

Figure 12.   Front-page of Otherside website.

metaverse game, similar to the virtual land NFTs in the Sandbox and Decentraland.

There will be a total of 200,000 Otherdeed plots in the Otherside metaverse, which will be dropped and sold in multiple drops. The 10,000 rare alien creatures "Koda" will also be randomly assigned to these plots.

The release of Otherdeed was a high point of the NFT wave, triggering an unprecedented rush and a Gas War on the Ethereum network, during which the Ethereum network gas exceeded a staggering 6,000 Gwei at one point.

Otherdeed NFT's trading volume also grew rapidly during the boom, reaching nearly $1 billion, with a total market cap of over $800 million.

Total transaction value: $997 million
Market capitalization (as of June 24): $850 million
Current floor price: 2.4 ETH
All-time high price (in ETH): Otherdeed #59906 sold for 625 ETH on May 8, 2022

5.4.   *Meebits*

Launched by Larva Labs in May 2021, Meebits is the third NFT series that are launched by the Larva Labs team (see Figure 13). The project currently ranks fifth in market capitalization at $800 million; however, it ranks first in total historical volume at $9.54 billion.

Figure 13.   Front-page of Meebits website.

Meebits are 20,000 unique 3D voxel characters (as shown in Figure 13) created by a custom generation algorithm and then issued based on the Ethereum blockchain. Owners of Meebit also have access to an additional asset package containing a full 3D model. You can use this 3D model to render and animate your Meebit, and you can animate Meebit to take any action you want, or use it as your avatar in the metaverse to put on your Facebook, Twitter, Instagram, or any other social profile. As a result, Meetbits has a much wider range of uses than CrytpoPunks, which is one of the reasons why it is so popular.

Meebits uses different features to represent the scarcity of individuals, including hair color, clothing, footwear, hats, glasses, accessories, etc. In addition, unlike CryptoPunks, the order in which they are minted also affects the value of Meebits, with earlier minted Meebits tending to be more valuable than later ones.

On March 12, 2022, Yuga Labs acquired Meebits along with CryptoPunks, and the acquisition had a positive impact on Meebits, with its floor price rising from 3.2 ETH on March 9 to 6.8 ETH on March 12.

Total transaction value: $9.54 billion
Market capitalization (as of June 24, 2022): $806 million
Current floor price: 4.95 ETH
All-time high price (in ETH): Meebit #4092 sold for 15,200 ETH on January 19, 2022

5.5. *Art Blocks*

Art Blocks is an NFT project platform for programmable random generation of encrypted artwork created by Erick Snowfro in 2020 (as shown in Figure 14). Thus, unlike the previously mentioned NFT series, Art Blocks is an NFT sub-project platform of multiple art styles led by many well-known artists, focusing on truly programmable on-demand generated content that is immutably stored on the Ethereum blockchain.

Designed to host projects by innovative digital artists, Art Blocks combines creative coding with blockchain technology to create a new paradigm for art creation and ownership. Collectors actively participate in the realization of the artist's vision through the generation of unique algorithmic artworks. This symbiotic relationship and shared experience form the basis of a vibrant community.

Works on the Art Blocks platform are submitted by individual artists or partners and are approved by the platform's curatorial board prior to going live. These highly innovative works by renowned artists are artistically and technically stunning and innovative, and the Art Blocks platform has already begun to see some headline-grabbing effects. For example, there are some huge-volume series such as Snowfro's own Chromie Squiggles series (the first series to be shelved on Art Blocks, with a trading volume of 17,100 ETH), the Fidenza series by Tyler Hobbs (trading volume of 47,500 ETH), and Dmitri Cherniak's Ringers series (trading volume of 26,800 ETH).

Figure 14. Front-page of Art Blocks website.

Total trading volume: $1.27 billion

Market capitalization (as of June 24): $840 million

Current floor price: 0.029 ETH

All-time high price (in ETH): October 2, 2021, Ringers #109 sold for 2,100 ETH.

5.6. *Moonbirds*

Moonbirds is a pixelated collection of owl character NFTs created by tech entrepreneur Kevin Rose's PROOF Collective (as shown in Figure 15).

Moonbirds feature a unique PFP design that allows them to be locked and nested without leaving your wallet. Once your Moonbirds

Figure 15. Front-page of Moonbirds website.

nest, they begin to reap additional benefits. As the total nesting time builds up, you will see your Moonbird reach new levels, upgrading their nest. Upgraded nests enhance air drops and rewards.

Moonbird holders have access to the program's private PROOF Discord, where they can get access to a multitude of information, including exclusive access to Moonbird-related drops, parties and IRL events, as well as access to upcoming PROOF projects and access to the PROOF metaverse known as Highrise.

MoonBirds sold out quickly after casting opened on April 16, 2022, sparking a community chase that generated over $300 million in sales in just five days.

MoonBirds' success is due to the strong operations of PROOF Collective, an NFT-based membership club with an experienced team whose founders, Kevin Rose, Ryan Carson, and Justin Mezzell, are all well-known figures in the Web2 and Web3 fields.

Kevin Rose is a partner at venture capital firm True Ventures, founder of social news site DIGG, and an angel investor who has invested in well-known tech giants like Twitter, Facebook, and Square. From 2012 to 2015, he was a general partner in Google's venture capital division. He has also hosted a number of technology and Crypto-related podcasts, and he created the PROOF podcast that led to what became the Proof Collective Club.

The Proof Collective Club is a group of 1,000 PROOF Collective NFT holders that includes well-known NFT artists Beeple, Sabastien Borget of the Sandbox, Erick Snowfro of Art Block, and investor Gary Vaynerchuk, among others.

Total transaction volume: $529 million
Market capitalization (as of June 24, 2022): $430 million
Current floor price: 19.4 ETH
All-time high price (in ETH): Moonbirds #2642 sold for 350 ETH on April 23, 2022

5.7. *Loot*

Loot for Adventurers is an on-chain NFT with a black background that contains only text and allows anyone to participate in casting for free during the casting period. Loot is an adventurer kit randomly generated and stored on-chain that intentionally omits stats, images, and other features for others to interpret. You are free to use Loot in any way you want.

The total number of Loots is 8,000, and each Loot represents an Equipment Pack. Each equipment pack contains eight pieces of equipment, consisting of eight units of variables, where the concept of game equipment is utilized: weapon, chest armor, helmet, belt, shoes, gauntlets, necklace, and ring. The interpretation of the various combinations is entirely up to the community, giving the holder great room for re-creation, making Loot a programmable NFT. The successful launch of Loot saw the emergence of a large number of copycat disk projects, including More Loot, pLoot, Rarity, Bloot, etc., but there is no shortage of scam projects among them.

Loot has transformed the way NFT tokens are created,[3] from passive acceptance of tokens created by project owners to community co-creation of tokens; therefore, Loot is seen as a new paradigm for NFT products.

In addition, Loot founder Dom Hofmann, who co-founded Vine, a short video sharing App owned by Twitter, as well as the NFT community Blitmap and the video game project Supdrive, is one of the reasons why Loot has gained so much attention.

Total trading volume: $1 billion
Market capitalization (as of June 24): $340 million
Current floor price: 1.3 ETH
All-time high price (in ETH terms): Loot (for Adventurers) #2579 sold for 15,180 ETH on January 19, 2022

6.   Conclusion: The Future of NFT

As of June 27, 2022, according to NFT.GO data, the total NFT market has reached nearly $60 billion in transaction volume, with

---

[3]Why is Loot Project Trending in the NFT Gaming Community? Binance Academy. https://academy.binance.com/en/articles/why-is-loot-project-trending-in-the-nft-gaming-community.

the past year alone accounting for $58.8 billion of that volume. NFT development has entered its first peak.

NFT is changing our lives, impacting the way we socialize, create, buy and sell art, music, video, images, brands, and more. NFT is rewriting our business models, integrating NFT capabilities with social products like Twitter and Instagram, which have hundreds of millions of users. YouTube is also exploring NFT capabilities for its video creators. Reddit has already launched its own line of NFT favorites.

The future of NFT has to evolve in conjunction with other applications. For NFT to reach its great potential, it must integrate with other applications such as metaverse, games, DAO, DeFi, etc. to leverage its unique nature and expand its presence in a variety of industries.

NFT, as an asset class, is a natural fit with the metaverse. The most practical, easy, and effective way for a brand to explore the metaverse universe is to start with new marketing of NFT digital collections, from sports leagues such as NBA, Premier League, and La Liga, restaurant giants such as Coca-Cola and McDonald's, luxury brands such as LV, Gucci, and Burberry, to Ubisoft, Square Enix, and other game majors, all are jumping on the bandwagon to explore and apply NFT.

Many metaverse worlds are built based on NFT, such as the Sandbox metaverse world, all the buildings and facilities of this world need to be built on its NFT land. In addition, NFT has unique characteristics that allow it to act as the user's unique identity in the metaverse.

As a result, the metaverse and NFT will become more and more closely integrated, and NFT will become an integral part of the metaverse.

NFT may propel blockchain gaming to become a trillion-dollar industry. The crypto bull market has brought a surge in tokens, prompting the Play-to-Earn (P2E) model to spread like wildfire among blockchain games, which has also ushered in the climax of development. From CryptoCat, Axie Infinity to StepN, they have all generated billions of dollars in profits for players, and the foundation of their success comes from the NFT element of play, which is the soul of their products. Year 2022 has seen P2E chain games become one of the biggest trends in NFT popularity.

The combination of DAO and NFT has brought about an evolution in organizational governance models. From collective ownership of NFTs to NFT community governance, DAOs bring many benefits to the development of NFTs. If an NFT is too expensive, then the DAO can attach collective ownership to the NFT, allowing all DAO members to share the NFT in common and share the benefits that the NFT brings.

New NFTs created by celebrity artists always sell out and can raise funds for their new works easily, but for newcomers to the art-making field, their work may hardly be seen. The DAO-run NFT Creators Collective helps art creators raise money, market their work, and more to drive the NFT creator economy.

In short, NFT is moving up and changing the business model in both the physical and digital worlds. As this technology advances, NFT will become part of our future lives, and NFT will accelerate and become more prevalent globally in the future. NFT is the future of the metaverse and crypto space. However, NFT is still in its early infancy, and the future of NFT will see many changes, perhaps far from the NFT we see today, and hopefully, we will see many more new uses for NFT in the coming years. What is clear is that more and more NFT opportunities are emerging. I hope you are able to step into this NFT wave.

References

Guan, C., Ding, D., Guo, J., & Teng, Y. (2023). An ecosystem approach to Web3.0: A systematic review and research agenda. *Journal of Electronic Business & Digital Economics*, 2(1), 139–156.

Heidorn, C. The 3 best NFT aggregator platforms in 2022. https://tokenizedhq.com/nft-aggregator/.

Lee, M. More than JPEGs: How bored Ape Yacht club built an NFT empire. https://www.nansen.ai/research/more-than-jpegs-how-bored-ape-yacht-club-built-an-nft-empire.

Non-fungible Token (NFT). (2022). Market research report. Virtue Market Research. https://virtuemarketresearch.com/report/non-fungible-token-nft-market.

Chapter 6

Blockchain Gaming

Jiancang Guo

According to research reports, the global gaming industry reached $175.8 billion in revenue in 2021, driven by nearly three billion players worldwide, and this figure is expected to grow to more than $200 billion by 2024, and venture capitalists like a16z with a keen sense of the market are turning their investment focus on the metaverse gaming space.

As one of the core infrastructures of the metaverse, blockchain naturally has a very close relationship with metaverse games.

Statistics show that in the first quarter of 2022, blockchain game projects have raised more than $2.5 billion from investment institutions, and according to the data, gaming has become the most popular blockchain investment track after public chains.

Nonetheless, there has been a great deal of controversy about the combination of blockchain and gaming, especially since the current play-to-earn model has proven to be unsustainable and games that lean too much toward finance (GameFi) have little appeal to traditional gamers.

So is there a future for blockchain games? This chapter attempts to generate some clues that can help you answer this question from a gamer's perspective.

1. Why Do We Play and Why Do We Give Up Games?

When we choose to play a game, there are inevitably some reasons that drive us to explore the game, and as the short- or long-term gaming experience passes, we choose to give up a game for a number of reasons.

For example, Vitalik Buterin, the co-founder of Ethereum, was a long-time player of World of Warcraft, but he finally gave up the game he played for three years because Blizzard developers nerfed his favorite warlock spell skill. Another example is that some people who were also veteran players of World of Warcraft eventually chose to go AFK because of the hyperinflation of the currency in the game due to the new version upgrade.

So what are the motivations for us to play a game? The following are four of the more common reasons:

i. It is recommended by friends; or people around them are playing, so they want to play the game with them.
ii. We are purely attracted by the quality and gameplay of the game.
iii. We look for some entertainment to pass the time.
iv. We use it as a tool to earn a living (to make money).

In the field of traditional games, the first three types of gamers account for the vast majority of the games played, while gamers who play to make money only account for a very small portion (some are gold studios and some are professional-level players).

There are more reasons for players to give up playing a game and the following are some of the common ones:

i. The game itself is not playable.
ii. We played for too long and lost the sense of freshness.
iii. The people around you are not playing and are too boring to play by themselves.
iv. The monetary spending in the game becomes too high, and the investment is not proportional to the returns.
v. The game lost balance because of external forces (external script).

vi. After starting a career or having a family, there is not much time to play.

vii. The game is not making money for the player.

viii. Due to the official unilateral changes of the game, resulting in negative feelings about the game.

Therefore, in theory, if there is a game that can take into account all these considerations at the same time, then the possibility of it gaining long-term success is very high. However, unfortunately, the blockchain games that have emerged so far and are playable in the past have failed to fulfill the shortcomings mentioned above, and the vast majority of them focus too much on the financial part of the game (earn). So most of the existing blockchain games will eventually disappear in a short time.

In the following section, we will take a quick look at the history of blockchain games through a few representative blockchain games and then review some of the highly anticipated blockchain games in development and their respective characteristics to get a better understanding of the blockchain gaming industry.

2. Blockchain Game History

2.1. *The first generation of chain games: CryptoKitties and Decentraland*

The beginnings of the chain game can actually be traced back to the launch of Huntercoin in September 2013 (Newzoo, 2022), a peer-to-peer (P2P) decentralized cryptocurrency game that forked the Namecoin blockchain (Figure 1). Huntercoin allowed players to "mine" cryptocurrencies in the game world. However, unfortunately, the game's founder, Mikhail Syndeev, passed away in February 2014, and the game faded from view.

The project that really caused the first chain game craze was CryptoKitties, which went live on November 28, 2017 (Figure 2). It was the first non-fungible token (NFT) project to use ERC-721, and the game initially issued 50,000 smart contract-generated crypto cat NFTs (also known as primordial cats), each with different attributes. Once players purchased the NFTs, they could begin a game of kitten

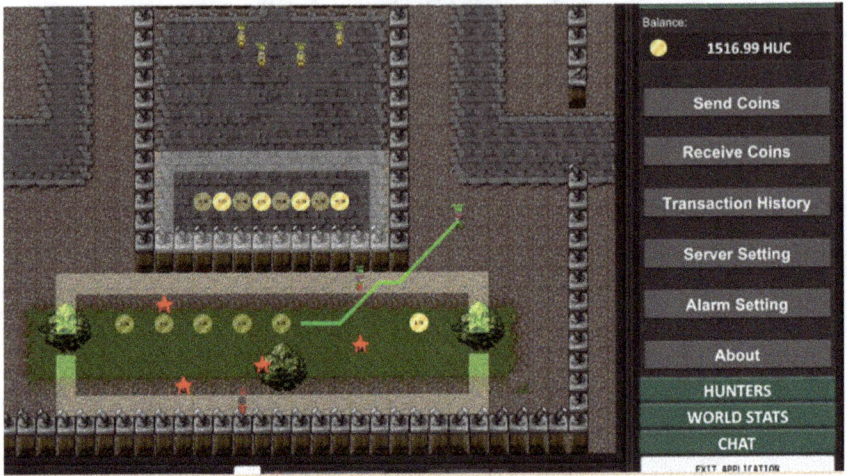

Figure 1.   Game Huntercoin.

Figure 2.   Game CryptoKitties.

breeding, where the genes of the hatched kittens were partly inherited from the previous generation and partly randomly generated.

The popularity of CryptoKitties has caused gas fees to soar for a while, and it didn't take long to successfully clog up the entire Ethereum network.

According to OpenSea's data, the CryptoKitties game now has more than 116,000 holder addresses and over 2 million crypto kittens (see Figure 3).

The success of CryptoKitties brought the team behind its development, Dapper Labs, on the investors' radar, making it one of the most competitive chain game teams at the moment.

In the same period, another noteworthy virtual universe game Decentraland was born. Since 2016, the Decentraland development team has been working on the Bronze Age version of the game, and in March 2017, Bronze Age was deployed on a test network. After testing, the team released a white paper outlining the vision for the Decentraland project, a virtual world subdivided into chunks of land with details stored on a blockchain ledger, which can be accessed using Decentraland's native token, MANA.

On August 17, 2017, Decentraland launched its first token offering, and the quota of 86,206 ETH was snapped up in just 35 seconds (Figure 4).

Figure 3. Game CryptoKitties's players.
Source: OpenSea.

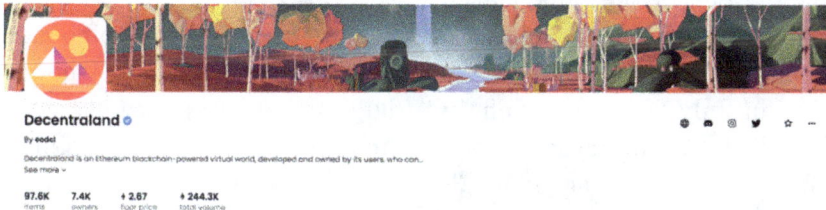

Figure 4. Front-page of Decentraland.

According to OpenSea's data, Decentraland currently has 7,400 landholder addresses, 97,600 virtual land lots, and over 244,000 ETH of historical transactions, making it one of the most successful virtual universe games to date.

However, in the Decentraland virtual world, players cannot find many attractive games or activities other than being able to visit the virtual world through avatars.

In general, the playability of the first generation of chain games was not very strong, and there was no special emphasis on the economic aspect of the game, so the reflexivity of the first generation of chain games was not that strong.

2.2. The explosion of GameFi: Axie Infinity and STEPN

CryptoKitties' success in the 2017 crypto bull market inspired Trung Thanh Nguyen, founder of Vietnam-based Sky Mavis, to form a team over the next year to implement his idea of the Axie Infinity game, which combines Pokémon gameplay with NFT and cloud-based cat farming.

On November 7, 2019, Axie Infinity developer Sky Mavis announced that it had raised $1.465 million in a funding round led by Animoca Brands.

Notably, Axie Infinity also introduces a new play-to-earn model for the first time. To participate in the game, users will first need to purchase Ether to get three Axie pets, which are purchased directly from players, and when marketplace transactions occur or players raise new Axie, the agreement takes a 4.25% commission. Unlike traditional models, in Axie's P2E model, 95% of the revenue goes to the player, not the game maker.

This novel model stimulated a large number of Filipino and Venezuelan players to participate in the game, and data showed that in July 2021, Axie Infinity's daily active users once exceeded 2 million, and its revenues for that month surpassed that of King's Glory, reaching a staggering $300 million.

Against the backdrop of this rapid growth, Sky Mavis, the developer behind Axie Infinity, closed a $152 million Series B round of funding (valued at \$3 billion) in October of the same year with participation from a16z, Paradigm, Accel, and other well-known investors.

However, this rapid growth did not last long, and Axie Infinity's monthly revenue started to fall off a cliff in December 2021. By March 2022, Axie Infinity's monthly revenue had fallen to \$2.2 million, down more than 99% from its peak, and to make matters worse, Ronin, the Ethereum sidechain on which Axie Infinity is based, was hacked, and some \$600 million in crypto assets were stolen.

Surprisingly, the theft of funds did not dissuade venture capitalists from continuing to be bullish on Axie Infinity. On April 6, 2022, Axie Infinity developer Sky Mavis announced a new round of \$150 million in funding, with participation from Binance, Animoca Brands, a16z, Paradigm, and other well-known investors.

According to Token Terminal statistics, Axie Infinity's monthly revenue has now fallen to \$200,000, down more than 90% from March 2022 when the assets were stolen (see Figure 5).

The play-to-earn model pioneered by Axie Infinity has attracted many imitators, but the vast majority of them have not been successful, with one exception being STEPN, a game built on the

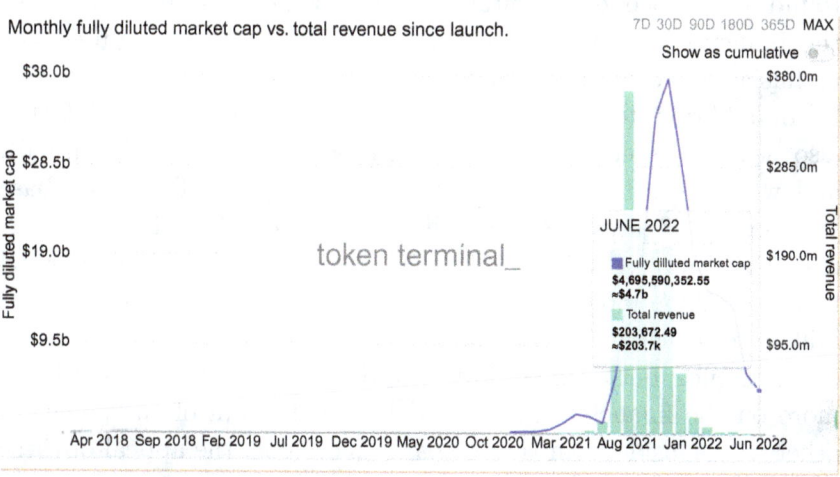

Figure 5.   Axie Infinity's monthly revenue.
Source: Token Terminal.

Figure 6.   Game STEPN.

Solana blockchain that borrows idea from Axie Infinity and replaces play-to-earn with move-to-earn.

In January 2022, STEPN announced a $5 million seed funding round led by Sequoia Capital India and Folius Ventures, and with the influence of Binance Launchpad, STEPN quickly ignited the enthusiasm of chain game players.

In addition to the improvement of economic model, STEPN also seized the entry point of "exercise and health" (Figure 6), and while Axie Infinity's revenue was sliding, STEPN's user base and revenue figures were climbing rapidly. In May 2022, STEPN co-founder Jerry Huang revealed in a media interview that STEPN had 2–3 million monthly active users worldwide and generated $100 million in revenue that month.

In essence, both Axie Infinity and STEPN are projects that focus more on the financial aspects and less on the fun of the game. In other words, aside from the "to-earn" mechanic, the appeal of these games to users is very low.

This also explains why these X-to-earn games are very explosive, but the ensuing reflexive impact can also be very strong.

3.   The Next Phase of Blockchain Games

So what if the game itself is fun and it engages real players? That might be the next phase of blockchain gaming.

In the past year, apart from investing in existing chain games, investment institutions have also started to lay out some high-quality chain game projects, and some of the top crypto institutions have participated in investing in some so-called 3A chain game projects.

For example, a16z invested in Forte, Yuga Labs, and Improbable, Paradigm invested in Parallel; Binance Labs invested in Ultiverse; and FTX (Alameda Research) invested in Yuga Labs, Big Time, and Star Atlas. In addition, Ubisoft, the world's leading developer of 3A games, has also started backing some chain game projects, such as Frontier Game.

Next, let us take a brief look at some of the high-quality chain game projects that are progressing relatively quickly.

3.1.   *The emergence of "3A chain game"*

(a) Illuvium

On March 12, 2021, Illuvium announced a $5 million seed funding round led by Framework Ventures, followed by a liquidity bootstrap pool (LBP) funding campaign in late March for the governance token ILV, which has three main functions in the ecosystem:

  i. governance;
 ii. liquidity mining rewards;
iii. pledge for game dividends.

Illuvium is an AAA-rated crypto game developer and publisher co-founded by CEO Kieran Warwick, game designer Aaron Warwick, and art director Grant Warwick, and currently employs around 100 people, according to information from LinkedIn. According to the plan, Illuvium intends to develop several games for its issued NFT assets and tokens to create a metaverse.

After about one year of development, Illuvium started the beta phase of Illuvium. It is a game in the form of a self-moving strategy in which players place NFT characters on a grid-like battlefield during

Figure 7.   Game Illuvium.

the preparation phase and then automatically fight each other's characters (Figure 7).

Later Illuvium raised over $72 million more in the first land sales between June 2 and June 5, 2022, for only one-fifth of all the land resources needed to extract the three key fuels from Illuvium lands for Illuvium games.

Related to these land NFTs is Illuvium's Overworld open-ended gameplay. During the NFT.NTC conference in June 2022, Illuvium released an official Overworld gameplay trailer that allows players to explore the game world, harvest game resources, and collect NFT creatures.

Looking at the completion of the roadmap, Illuvium's self-propelled battle arena game is the most completed, and the Illuvium Zero stage (a mini Overworld gameplay) is expected to be launched shortly, in which the land will yield resources that can interact with the main game (Figure 8).

The final Overworld open-ended game, on the other hand, may require players to wait a very long time.

Illuvium ✔ @illuviumio · 6月22日   ...
Illuvium: Overworld Gameplay Footage Release - Work in Progress

(probably nothing 😉)

▶ 4.8万 次观看   5:29 / 8:18 🔊 ↗

💬 106   🔁 997   ♡ 1,771   ⬆

Figure 8.   New Game Illuvium Zero.

(b) Parallel

Parallel, a sci-fi NFT trading card game (TCG), built by game development company Parallel Studios, has received $50 million in funding led by Paradigm, with other notable investors include YouTube co-founder Chad Hurley and Polygon.

Under their plan, Parallel will have six NFT pack drop events where cards will be sold every other quarter, and these NFT cards can be used to form sets belonging to five factions and to compete against players using these sets. The winner wins Echelon Eco's Parallel token rewards[1] (total fixed at 111 million) and experience (XP), and players can consume Parallel tokens as well as expcrience (XP) to

---

[1]Echelon Prime Foundation. Echelon White Paper. https://paper.echelon.io/echelon-whitepaper.pdf.

upgrade or inherit a set of cards. Inherited cards will have a cool-down period to prevent the number of cards from growing too quickly, while upgraded cards can reduce the cool-down period and unlock advanced skins without affecting the balance of the game.

For traditional players with a low-risk appetite, Parallel offers inexpensive non-NFT card packs that have the same use cases as other NFT decks, but significantly reduce the PRIME token rewards of the P2E system.

Unlike the dual-token model adopted by other blockchain games, Parallel uses a single-token model (i.e., its governance tokens also act as functional in-game tokens), and the PRIME tokens consumed by its games do not leave circulation but are reallocated to different pools to reward different pledgers.

Like other ambitious game publishers, the TCG card game (Figure 9) is the first game Parallel Studios plans to build, in addition to a library of 3D models and AR models for NFT cards that can be used directly in the development of future first-person shooters (FPS) and role-playing games (RPG).

Notably, Parallel also announced a partnership with Superconductor, the company founded by the Russo Brothers (directors of

Figure 9.   TCG card game.

several other Marvel movies including "Avengers: Infinity War"), which could hint at something ambitious.

As of now, the Parallel TCG card game is under internal testing by the team and Mod managers, and according to projections, the game is expected to officially launch in early 2023 when PRIME tokens are open for circulation. The other games in the ecosystem are still in the preliminary stage of material library preparation.

(c) Phantom galaxies

Phantom Galaxies, an open space simulation game created by Animoca Brands and game developer Blowfish Studios, will support multiple crypto communities through a multi-chain framework and is scheduled to debut on the Polygon blockchain.

According to the whitepaper (Ultra Modern Group Ltd., 2021), Phantom Galaxies will have a range of NFTs and fungible tokens (tokens) for governance, starfighter fusion, and access to certain activities.

In the game, user-owned land comes in the form of Planets and Asteroids (Figure 10). Phantom Galaxies received a $19.3 million investment from Sequoia China, Polygon Ventures, Dapper Labs, and YGG in a planetary private funding campaign in May 2022. Each planet was assigned a certain number of System Governance Tokens

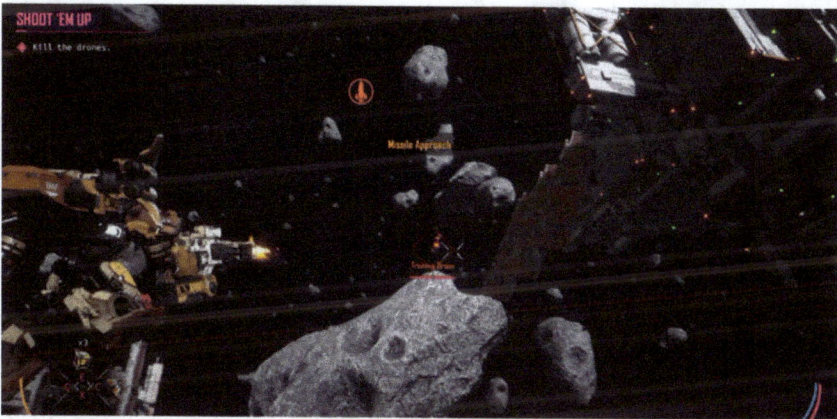

Figure 10.   Phantom Galaxies.

(with a set minimum and maximum) to encourage participation in the game experience, and the total number of System Governance Tokens was capped at a fixed value of 888,077,888.

In turn, in-game resources and materials will be represented by an uncapped number of fungible tokens that can be harvested and scavenged in the open universe and then traded on the Phantom Galaxies' marketplace.

According to the roadmap, Phantom Galaxies has opened for early gameplay in late 2022, with official pre-beta gameplay footage released, as shown in Figure 10.

Due to limited space, this chapter only briefly mentions the design and progress of the three new chain games. In addition to the different types of games, the economic model design of these blockchain games is also very different. For example, Illuvium and Phantom Galaxies both adopt a dual-token model, separating governance tokens from in-game economic tokens, and most of the game's output is based on economic tokens, supplemented by governance tokens. This model follows Axie Infinity's experience of shifting the token dump from new output to the game's economy tokens, while governance tokens are easier to speculate on. However, the downside of the dual-token model is that the use of governance tokens is very limited, and teams that take the dual-token route will eventually run into this problem unless they find a way to increase the governance token's value or utility; otherwise, it will be abandoned and many speculators will be left out of pocket.

Parallel, on the other hand, takes a single-token model, where the reward players get in the game is an ecological governance token cum game utility token. The single-token model will be more of a test for the game team because the single-token system increases the influence of the secondary market on the in-game economic system, and most single-token model chain games on the market have not been successful. This model requires to ensure that the game is really playable and needs to provide scenarios that consume tokens; otherwise a token dump would be very detrimental to the overall game development. To mitigate this issue, the low cost of non-NFT cards offered by Parallel will be very important to attract non-crypto-ecological gamers to the system.

3.2. *Some thoughts: Is play-and-earn viable?*

The previous generation of chain games, represented by Axie Infinity and STEPN, focused on financial games, and once these games collapse, the number of participating players will plummet, creating a negative feedback loop.

The new generation of chain games is focusing on the playability of the game itself, so they use the term "Play and Earn" to describe these chain games.[2] According to some test players, the playability of some Play-and-Earn games has surpassed that of traditional games in the same genre, which is probably a good sign.

However, as Eva Wu (2022) of Mechanism Capital noted, the "Play and Earn" tagline still attracts too many economic gamers rather than actual gamers. When gamers are told to "play" and then "earn," it sets an unrealistic expectation, which can lead to constant selling pressure on the game's native tokens.

A better model would be to keep gamers and financial gamers separate, with few intersections between them, and try not to use the tagline "Play and Earn" to get more players who care about the game itself to participate in the system (as shown in Figure 11).

By doing so, the whole game can run healthily instead of dying quickly.

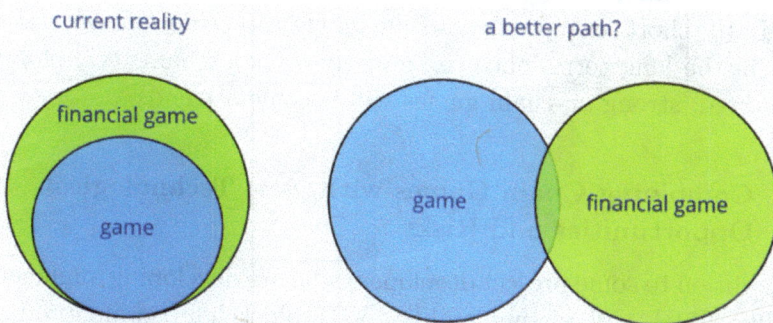

Figure 11.   Two models of play and earn.
Source: Eva Wu.

---

[2]Thoughts on on-chain gaming. https://dialectic.ch/editorial/thoughts-on-chain-gaming.

4.　Weak On-Chain Gaming vs Strong On-Chain Gaming: The Trade-Offs of Blockchain Gaming

Due to the limitations of the underlying layers of blockchain technology, the vast majority of current blockchain games are in the form of hybrids, where the chain game puts the game's governance tokens and core NFT assets on the chain, but the rest of the game runs on proprietary servers off-chain, which are called weakly on-chain games.

In general, weakly on-chain games offer a better experience, better game quality, and are more likely to attract non-crypto players. However, the downside is that the game publisher plays a central authoritative role and can unilaterally censor, change the rules, or disappear, in addition to the interoperability of weakly on-chain games being an issue.

Games that deploy the entire game on the chain are called strongly on-chain games.

For example, Dark Forest, Dope Wars, Briq, Loots, and The Realms all belong to strong on-chain games, and the current strong on-chain games still have a large disadvantage in game experience and quality, and generally attract native crypto gamers.

However, these games also have their own advantages, such as trust minimization, open source, potential composability and interoperability, and no licensing, among others.

In the short term, weak on-chain games can achieve faster scaling, but in the long term, once the underlying blockchain technology is perfected, strong on-chain games will become a trend.

5.　Combining Chain Games with New Technologies: Opportunities and Risks

In addition to some project developers who are developing augmented reality (AR) and virtual reality (VR) blockchain game products to enrich players' experience, there are also some teams who have started to apply artificial intelligence (AI) technology to blockchain games. For example, AI Arena tries to apply AI technology to fighting games, where players train their NFT fighters through training programs, and the fighters are powered by AI in the sparring.

Figure 12.   Scenes from the TV series Westworld.

If AI technology reaches a certain stage of development, we may also see scenes from the TV series Westworld (Figure 12), where it will be difficult to distinguish between gamers and AI NPCs.

Naturally, the development of new technologies will also impact the gaming industry. For example, as early as in March 2016, the AI robot "AlphaGo" developed by the US company Google has already defeated the human Go champion. In June 2018, the AI gaming team developed by OpenAI defeated the vast majority of human Dota 2 players in 5v5 battles. Tesla founder Elon Musk, a supporter of OpenAI, commented, "For the first time, OpenAI is beating the world's best players in a competitive e-sports game that is far more complex than traditional board games like Go or chess. It's far more complex".

When games are combined with Earn mechanics, AI technology will become a disruptive force, meaning that simple chain games will become "cash machines" for AI bots.

Magic: The Gathering is Turing Complete

Alex Churchill
Independent Researcher
Cambridge, United Kingdom
alex.churchill@cantab.net

Stella Biderman
Georgia Institute of Technology
Atlanta, United States of America
stellabiderman@gatech.edu

Austin Herrick
University of Pennsylvania
Philadelphia, United States of America
aherrick@wharton.upenn.edu

Figure 13.   A research paper by Cornell University.

And only complex games are likely to withstand the onslaught of AI power, as AI developers from Cornell University (Figure 13) reluctantly admitted in a 2019 research paper (Churchill *et al.*, 2019) that "the game's set of structures identifies Marvel as the most computationally complex reality game known".

With more than 20,000 cards and nearly 100 unique mechanics, this game means that it has reached the upper limit of computational complexity in terms of rules complexity.

Therefore, "complexity" could be a moat for blockchain games with financial incentives.

6.   Conclusion

Blockchain games have gone through two generations of development, and two to three masterpieces have emerged in each cycle. In the next crypto cycle, blockchain games that focus on gameplay are more likely to stand out in the fierce chain game competition, and besides gameplay itself, we also need to pay attention to the impact of emerging technologies such as AR and AI on blockchain games, because simplicity may not be a good feature for blockchain games.

References

Churchill, A., Biderman, S., & Herrick, A. (2019). Magic: The gathering is turing complete, p. 8. https://arxiv.org/pdf/1904.09828.pdf.

Newzoo. (2022). Global Games Market Report 2022. https://newzoo.com/insights/trend-reports/newzoo-global-games-market-report-2022-free-version.

Ultra Modern Group Ltd. (2021). Phantom Galaxies™ (PG) litepaper. https://phantomgalaxies.com/BFS%20Phantom%20Galaxies%20Litepaper%20v1.0.pdf.

Wu, E. (2022). Breaking the trance: Crypto gaming. mechanism capital. https://www.mechanism.capital/breaking-the-trance-crypto-gaming/.

Chapter 7

Decentralized Autonomous Organization (DAO): From Layers to Networks

Sue Tang, Bosheng Ding, and Jack Cheng

In 2005, in his book *The World is Flat*, Thomas Friedman claimed that the emergence of the Internet was rapidly flattening the world (Freidman, 2005). The pyramid-shaped hierarchy is gradually dwindling as compared to organizations in the first half of the 20th century. Control is waning and innovation is taking over as the primary driver of productivity, which in turn encourages more hierarchy reductions. However, Web2's capabilities are constrained, and technology firms that manage Internet access have become new monopolies due to the Matthew effect. Fewer companies can develop into new industry titans and it seems the pace of innovation has slowed down recently.

Throughout history, innovation has frequently slowed down. However, we cannot halt history, just as we cannot stop innovation. DAO (Decentralized Autonomous Organization), a new flat-network organization paradigm, has started to draw interest, particularly in the wake of blockchain technology. A free, open, egalitarian, and decentralized organization is what DAO stands for conceptually. Web3 requires a complementary organizational structure in order to support innovative, collaborative output. The DAO is still impossible to describe and anticipate, but its development appears to support Friedman's assertion from nearly two decades ago that the world is turning into a web.

1. The Rise of DAOs

DAOs may have existed for as long as there have been people on earth. The organizational form of humans was self-organization with dozens or hundreds of individuals as units, involving leader election, a referendum on key events, transparency, and openness. This organizational form prevailed in regions where the prehistoric living environment was not harsh. When the environment changed, humans developed a complex hierarchical society in order to control floods, protect against invading foes, or raise money for commerce. And, for the first time, human society's production surpassed its consumption after multiple industrial revolutions. On the one hand, the lessened pressure for survival creates greater room for creativity; on the other hand, it also leads to a change in organizational structure.

The first stock option plan in history, created and implemented by the American Fizel Corporation in 1952, gave employees a stake in the company. Today, many businesses have implemented equitable incentives. The idea of a DAC (Decentralized Autonomous Company) was presented by Daniel Larimer, the founder of EOS, on September 7, 2013 in his blog, "Overpaying for Security" (Larimer, 2013). A few months later, the idea of the DAO as we know it today was formally introduced by Vitalik, the creator of ETH (Buterin, 2013).

In 2016, the first DAO was created — "The DAO" — a venture capital fund for crowdfunding, founded on ETH by the blockchain business Slock.it. This was the DAO's initial experiment as well. However, after a hacking event, ETH underwent a hard fork. All blockchain initiatives started to look for new directions with the DAO following the hacking. Through a number of DAO initiatives, including Uniswap, Bankless, Mirror, ConstitutionDAO, Cosmos, and SeeDAO, individuals started making progress toward the equality, freedom, and transparency that the DAO advocates, as do teams and companies.

2. DAO's *Status Quo*

DAOs in 2021 can be characterized as voluntary groups that follow the tenets of digital cooperation, according to the essay "A Prehistory of DAOs" (Kreutler, 2021). They are anonymous voluntary groups that bring together strangers, friends, and allies in the name of a common cause, supported by a token mechanism, rewards, and governance. There are currently 4,833 DAOs on the market, with an asset management scale of 8.2 billion, according to Deep DAO data (as of 12:00 p.m. on June 22). There are 676,200 active users and 3.7 million governance token owners of DAOs (see Figure 1).

Along with the ongoing expansion of the development scale, the ecological activity within the DAO has also done well. A total of 61,200 suggestions and choices have been made, and 3.6 million votes have been cast overall. The highest level of active proposal voting was seen in November 2021 (see Figure 2).

"If 2020 was all about DeFi, and 2021 was all about NFTs, 2022 will be the year of the DAO" said Messari in the report "Crypto Theses for 2022" (Selkis, 2021). The DAO's ecosystem is definitely expanding quickly, and there are several new industry sectors that are continually adopting DAO-based practices. The DAOs that are now available on the market can be loosely categorized into the following groups, according to Cooper Turley's compilation: Protocol

Figure 1.   Deep DAO.
Source: deepdao.io.

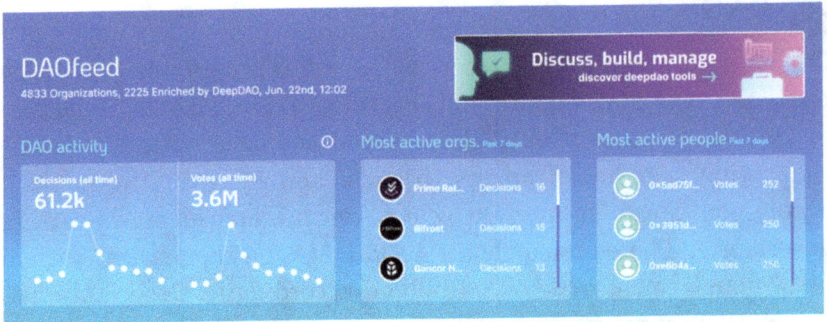

Figure 2.   DAOfeed.
Source: deepdao.io.

Figure 3.   Categorization of DAOs.
Source: DAOs: Absorbing the Internet.

DAOs, Service DAOs, Social DAOs, Media DAOs, Investment DAOs, Funding DAOs, Creator DAOs, and Collector DAOs (Turley, 2021) (see Figure 3).

Grant DAOs: A DAO's primary practical use is financing, much like angel investors. These DAOs serve as a sort of community motivation and are linked to already-existing enterprises. The community contributes money and uses the DAO to vote on suggestions for governance that would determine how the money is split up among the various contributors. Grants DAOs aims to develop the larger ecosystem, aid innovative initiatives, and create new pathways for Web3 participants. The well-known GitcoinDAO and MolochDAO, a forerunner of DAOs, are among the more well-known Grants DAOs.

Protocol DAOs: As their name implies, Protocol DAOs are collaborative organizations that exist to aid in the development of a protocol. Projects now have a new method of issuing tokens to the market thanks to Protocol DAOs. The creation of transferrable ERC20 tokens with secondary market value was also a first for Protocol DAO. The community and developers typically use these tokens to jointly control the system. All networks can issue tokens that are controlled and operated by their communities using the framework that Protocol DAOs provide (hopefully). An extremely common example of a protocol DAO is MakerDAO.

Service DAOs: In the Web3 industry, service DAOs serve as talent aggregators. Many real-world freelancers are also involved in DAO-like organizations, not just in Web3, as Cooper Turley put it in "DAO Landscape". From legal to creative, governance to marketing, and development to capital management, service DAOs have established a route for contracting Web3 digital nomads. Typical service DAOs include Party DAO, DeepDAO, Yam DAO, and the well-known gaming guild YGG.

Investment DAO: The DAO industry has a highly unique area devoted to investing in DAOs. The presence of investing in DAOs is about returns and profitability itself, unlike a majority of DAOs now available on the market that are in a non-profit state and persistently seek a profit model. Investing in DAOs aims to attract funding and backers for implementation. DAO investment will resemble closed-end private equity funds in several ways, except that this fund's investors and managers are distributed throughout the world. The LAO is a pioneer in the field of investment DAOs and was founded by Aaron Wright. In addition, investment DAOs like FlamingoDAO, BitDAO, and MetaCartel are worthy of note.

Media DAOs: The organizational structure of traditional media is single/linear. Generally speaking, it is the funding source controls that the organization's content, not its target audience (see Figure 4).

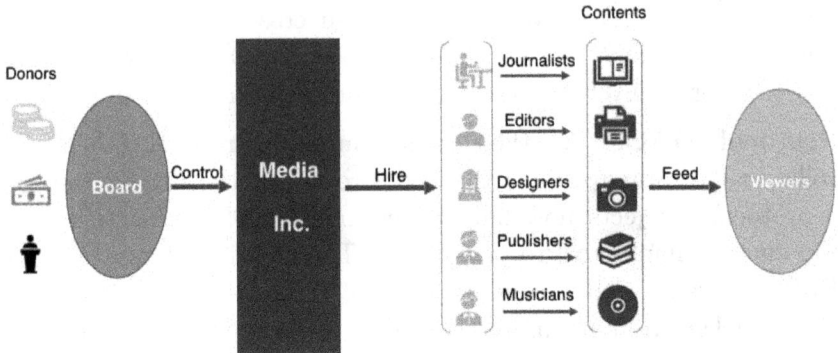

Figure 4. Traditional media structure.
Source: DAOrayaki: Designing a Decentralized Media Run by a DAO.

Decentralized media, commonly referred to as media DAOs, give control of the communication of information narratives back to people who have paid for the content. Media DAOs provide publicly available material that is written by individuals or groups, and the benefits of that content are shared among the group. Stakeholders can also participate in topic selection and resource management processes. By allowing readers and consumers to participate in the administration and direction-setting of media content, Media DAO alters the dynamic between artists and audiences. "Media DAOs have turned money for content into a two-way conduit", says Cooper Turley. BanklessDAO is without a doubt one media DAO platform that is now having the most success. ForeFront and DarkstarDAO are equally deserving of praise.

Social DAOs: Social DAOs resemble online gatherings, fan clubs, and interest groups in many ways. In addition, a lot of social DAOs contain requirements, making them more like a walled garden (such the requirement of holding a specific percentage of the community's tokens). Social DAOs like FWB and SEEDCLUB are well known.

Collector DAOs: Investment and collector DAOs have a similar appearance; however, collector DAOs place more emphasis on collecting (NFT) assets. There are numerous well-known collector DAOs, such as WHALE, NounsDAO, PleasrDAO, and NounsDAO.

Members of the collector DAO community cherish the long-term worth of the collection and the creator; hence, they typically do not sell the collections they acquire. This makes it different from investing in DAOs. Some collector DAOs will provide some curating and operations for the artists/creators in their communities in addition to taking part in the gathering of artworks/collections to aid in their improvement. There are numerous well-known collector DAOs, such as WHALE, NounsDAO, PleasrDAO, and NounsDAO.

Creator DAO: The creator DAO is primarily focused on people, whereas the media DAO is more concerned with publishing companies or collectives. The "fans" of particular creators have a way of supporting them through creator DAOs, which are more like fan economies. Additionally, through the creator DAO, individual creators have received the support they need, including financial assistance, to improve their ability to produce higher-quality works. This business operates on a win-win basis. Mirror is a well-known creator DAO.

3. Zero to One: Cold-Starting a DAO

Wang Chao, a deep DAO participant, once said in an AMA that cold-starting a DAO is very similar to cold-starting an Internet project. To set up a DAO from 0 to 1, the following conditions should be met.

3.1. *Set a clear goal*

Having a clear community goal or vision is a must for launching a DAO, which can attract people that share the same vision and quickly form an initial team. In addition, goals and visions can be iterated as the DAO evolves.

3.2. *Build a preliminary governance framework*

There are many DAO governance models. As each DAO has its unique goals and states, its corresponding governance model would also be different. However, building the preliminary governance framework should take three factors into consideration: ownership distribution, token weighting, and voting mechanism.

In a DAO, decentralization is not an end goal but an intermediate approach. The distribution of ownership needs to be decentralized so that community members can share the economic values, social values, resources, and governance components of the community to unleash their creativity. The current common ownership distribution method achieves distribution by issuing membership certificates such as governance tokens or NFTs. Token weighting is usually combined with a voting mechanism.

3.3. *Setting up the initial operating system*

The initial operating system here refers to an initial operation framework for the community, including member onboarding, guidance, participation, community contributions, and member compensations. The principle for setting up the initial operating system is to keep the entire system transparent and open. Employing some DAO collaboration tools, current ways to better set up the initial system include building a complete community manual, setting up bots, and setting up teams with different functions.

3.4. *A little "centralized" element*

Talking about "centralization" in a DAO might sound counterintuitive at first. However, a little bit of "centralization" is essential when starting a DAO in the first place.

When cold-starting a DAO in the early phase, even with a clear goal, it can be difficult to get everyone together to quickly build and run a complete DAO. On the one hand, the organizational structure of DAO is relatively loose. In the early phase, in order to quickly respond to complex problems, a core person or initial member is required to adopt a quick decision-making process. On the other hand, because of the freedom and threshold-free nature of the DAO, it is difficult to guarantee that all DAO members recruited in the early phase have the ability to coordinate and make decisions efficiently.

In "learning by DAOing: designing for the unique properties of DAOs", the authors found that while contributors want the ability to better shape strategies, they also do not want to be

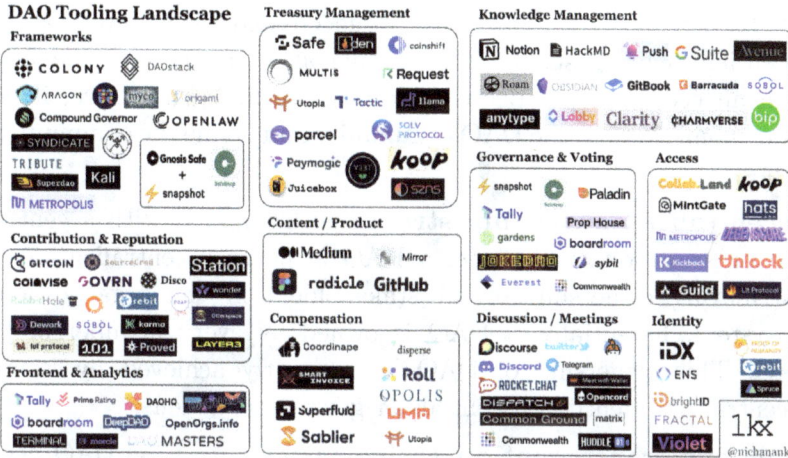

Figure 5. Organization legos: The state of DAO tooling.

handed a completely blank slate (Humenansky, 2022). The authors also found that the optimal balance is providing an initial charter outlining about 70% of what a working group will drive toward and empowering the community to provide feedback and fill in the rest. Therefore, in the initial stage of launching a DAO, it is necessary to establish a "centralized" core team. The DAO tooling landscape is presented in Figure 5.

4. DAO and Its Utopia

4.1. *Characteristics of an ideal DAO*

If DAOs are to serve as an innovative work organization paradigm that will innovate and interrupt existing organizational structures in the future, then ideally, DAOs should have the following characteristics:

Clear goals: A clear goal is not only a necessary condition to start a DAO but also an important feature of a DAO. A clear goal is like the North Star, providing directions on the value orientation and decision-making for the entire DAO.

Value creation: If a DAO cannot create value, the energy of its community organization will gradually weaken, and the DAO will become unsustainable. The value created by a DAO has the potential to bring sustainable funding and resources to the DAO from an economic perspective; from a cultural perspective, it can bring momentum to the DAO community and define the community's values.

Decentralized ownership structure: Compared with a centralized organizational structure, a DAO achieves decentralization by distributing ownership to all stakeholders in the organizational ecosystem. Essentially, a DAO is owned by every member of the DAO. This ownership of DAOs is currently achieved by issuing decentralized identity credentials or governance tokens.

Open and transparent operating system: An ideal DAO needs to have an open and transparent organizational operating system to maintain the openness, transparency, and traceability of documents and information within the organization and to ensure information symmetry among DAO members. Many blockchain protocols and various DAO management and collaboration tools emerging in the market are facilitating the achievement of this goal.

Autonomous and self-driven organization: A DAO hopes to achieve independent action at the organizational level without interference from other organizations outside the DAO. At the same time, its internal management should be bottom-up: DAO members are free and all DAO affairs are managed by its DAO members. In addition, DAO should be self-driven, and members of DAO should be able to participate in the construction and management of DAO spontaneously and create value for DAO.

Evolvable: An ideal DAO should not always stay in its comfort zone. Instead, it should continuously evolve with the changes in the environment, the structure of members, and other factors. It should constantly adapt to the new environment, evolve, and iterate continuously, so as to achieve the sustainability of a DAO.

4.2. *The ideal of DAO*

Any form of organization emerges in its social environment, and DAO is no exception. Let's take a look at its history: Guilds emerged in medieval Europe, companies emerged and developed from the Age of Discovery through the two industrial revolutions, and modern companies (employee shareholding with the flattened organization) emerged in the Internet age.

Fundamentally, relations of production depend on productive forces. But in recent years, for the first time in human history, we are about to enter an era of more production than consumption. According to the *Xinhua Daily* report, from 2018 to 2022, the total global grain production will be 2.65 billion tons, 2.71 billion tons, 2.78 billion tons, and 2.80 billion tons, respectively. In terms of numbers alone, it can basically meet the demand of 8 billion people in the world.

The productivity gains in the past have been driven by technological innovation. And now, it is almost impossible for humans to stop innovating and creating. Further innovation requires a more flexible social environment, a more open organizational structure, and more talented people in the world.

After satisfying the bottom two layers of Maslow's pyramid of needs (survival and safety), the productivity in society needs to meet higher-level human needs in the future.

Following this trend, we are making the world a better place. This is the ideal of DAOs and it is also the ideal of Web3.

5. What Do DAOs Bring to Ordinary People?

The emergence of DAO itself means that this is an era with an open and inclusive living environment. Therefore, according to the top three layers of Maslow's needs pyramid, what ordinary people need is belonging, esteem, and self-actualization. And, DAO has the ability to fulfill these needs.

5.1. *Sense of belonging*

DAO members can take part in the full development of DAO, bring up feasible proposals for independent communities, engage in the discussion of proposals through community meetings, vote on each proposal, create values for the community, participate in community management, and, ultimately, share every benefit of the community. To some extent, ownership means having a sense of security, naturally leading to a sense of belonging. In an ideal situation, this sense of belonging can also drive every member to actively contribute to the DAO.

5.2. Freedom and fairness

A DAO does not have an entry threshold and everyone can participate or withdraw freely. Therefore, community members are not controlled by the DAO. On the contrary, a DAO should belong to every member. On the one hand, a DAO's open and transparent governance mechanism means that its decision-making will be fairer, and everyone has the right to support or oppose every DAO proposal. On the other hand, a DAO is being regulated by its transparency and thus less likely to be immoral. To sum up, ideally, a DAO allows each participant to obtain a free and fair working environment.

5.3. Personal value and future

For members who join a DAO, they are initially attracted to the DAO by the common goals or vision of the DAO; members have the right to speak and vote for the projects they are interested in, and can really participate in the construction of the project. Meanwhile, a contribution made by an individual in the DAO will be rewarded by its application, including economic rewards (some DAOs will provide bounties or tips), governance power (such as governance tokens), and on-chain reputation (Community POAP). Because of the provable and traceable nature of blockchain, all these rewards are ultimately referred to as the on-chain resume/CV of each ordinary member, thus shaping his or her future in Web3.

6. DAO's Current Dilemma and Solutions

Although the DAO's ecology is booming and improving, DAO experiments are still facing difficulties and even failures; however, it is through trial and error that a DAO's governance can rapidly accumulate experience.

6.1. Token airdrop model: Contribution capture and centralization crisis

Most DAO ownership distribution is achieved by distributing governance tokens, and many DAOs distribute tokens at the beginning of their establishment to attract early participants. This may also be used as a form of their crowdfunding model. However, the current approach also presents two risks.

First, it makes a DAO's community potentially speculative. On the one hand, these DAO members attracted by the token will not contribute to the later community construction and the common goal of the community. On the other hand, some members who have contributed will feel that the situation is unfair. As a result, they will be dissatisfied with the community and become negative.

Second, this crowdfunding model may bring funds and resources to the DAO in the early stage. However, in the process of DAO development, those who bring capital and resources can control the large number of governance tokens obtained in the early stage. As a result, the development of DAO may cause an oligopolistic scenario, which will bring a crisis of centralization to the DAO.

The Juno airdrop in the Cosmos ecosystem is a good example. The rights and wrongs, the speculative nature, and inevitable ineffective centralization of the system itself reflect the failure of this farce. Therefore, subsequent airdrops began to absorb the experience of this simple airdrop model and added real contributions, such as the airdrop of \$OP's specific contribution behavior, which is an improvement (For details on the \$OP airdrop mechanism, please refer to the official website: https://www.optimism.io/). (See the Juno incident: https://www.defidaonews.com/article/6746316; https://polkachu.com/blogs/people-vs-juno-whale).

6.2. Balancing openness and stability: The problem of core contributors

For any organization, the change of core personnel is "severely traumatic". But because, in principle, members in the DAO can choose to join or leave at any time, compared to other organizations, this is more challenging for DAOs. For example, Andre Cronje's departure earlier this year has had a long-term negative impact on multiple projects such as Fantom and Yearn.

In addition to retaining the core contribution members of a DAO with ideals and interests like traditional organizations, emphasizing the following two aspects holds significant importance. (1) How to do a good job in a DAO's internal talent training management and talent welfare system, so as to better retain the DAO's core contributors. (2) The DAO should not rely too much on one or some contributors, and members within the DAO should be encouraged to participate in the construction of different fields.

An open system means that the system is in dynamic balance, so the DAO can theoretically have the ability to stay resilient and hedge the risk of losing any core members.

6.3.   Balancing collaboration and efficiency: Internal governance

Following the governance model of BTC, ETH, and other public chains, some DAOs (such as Uniswap, the proposer of the most simplified governance) need to go through a lengthy community vote, which greatly affects efficiency (see: otherinter.net). In order to solve the problem of how to balance efficiency and fairness at the same time, we are going to discuss two methods that some DAOs are exploring today.

Proxy voting mechanism: Representative democracy is chosen by more countries, which can balance fairness and efficiency to a certain extent. At Polkadot's Decode party in June 2022, the founder, Dr. Wood, proposed a wallet option that can select agents in governance voting. This undoubtedly refers to representative democracy.

Squads within the DAO: According to different dimensions such as member regions and different projects in the DAO, sub-DAOs of different sizes are hatched, the work is coordinated in a team mode, and the work content and project personnel are centralized, which can reduce decision-making costs and quickly improve work efficiency.

6.4. *DAO salary and welfare system: Source and distribution*

According to a CNBC article, "As inflation heats up, 64% of Americans are now living paycheck to paycheck". Globally, more than 684 million people live in poverty, not including those above the poverty line but struggling to make ends meet (Dickler, 2022). This means that if a future DAO is to achieve scalability and attract people from all walks of life to participate on a full-time basis, then the problem of compensation and the welfare system must be solved. At present, some DAOs have solved the problem of the return of their community contributors through the mode of bounty or tip while thinking about how to build a more reasonable salary system and DAO benefits, including reasonable vacation time, retirement plan, and members' psychological health consultation.

Regarding the establishment of the compensation and welfare system, we should think about the solution from two aspects: source and distribution.

First of all, the prerequisite for the establishment of the compensation and welfare system is that the DAO needs to establish a sufficient treasury. Therefore, exploring reasonable commercialization or seeking funding is feasible to build a DAO's treasury.

Second, the stages to distribute salary include the following: (1) In the early stage, each member's contribution can be recorded on the chain by other means, such as points, NFT or POAP, and other credentials. (2) After the distribution of a certain amount of funds, the initial bounty model, task mode, or tip mode can be used to make a reasonable return to members' contributions. (3) Ideally, after completing the above two stages, the DAO should set up a reasonable and feasible compensation and welfare system in the community (including leave system).

Of course, there is still a long way to go in the establishment of the compensation and benefit system for DAOs, and there is no perfect tool for implementation. However, only by solving this problem will DAOs be able to replace the existing company and evolve into the organizational model of the future.

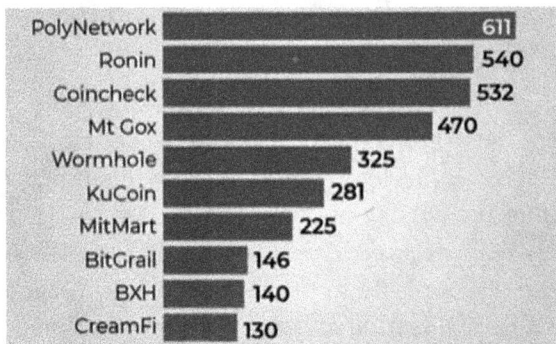

Figure 6. Money stolen by crypto hack (in millions of dollars).
Source: Bloomberg.

6.5. *The battle of light and shadow: The eternal hacker (and runner)*

Over the past two years, Web3 has seen several projects fail due to hacking. This seems like a curse since the first DAO. A DAO, like any other organization, has no reason to be fully protected. However, the improvement in the code is visible. For example, in 2021, DeFi on the BNB chain was attacked on a large scale. By 2022, however, the focus of the attack was on the cross-chain bridge. Another example is when the Dogecoin developer exited and other members of the community took over the project (this may not be a real rug pull).

In a sense, hackers are the watchdogs of every Web3 practitioner, constantly growing the industry in the most painful way possible (see Figure 6).

DAOs propose new possibilities for the development direction of the organizational model. For ordinary individuals, DAOs are a way to regain ownership and realize their values. The DAO was born for Web3.

References

Buterin, V. (2013). Bootstrapping a decentralized autonomous corporation: Part I. *Bitcoin Magazine*, 19 (2013).

Dickler, J. (2022). As inflation heats up, 64% of Americans are now living paycheck to paycheck. CNBC.

Freidman, T. (2005). *The World Is Flat,* vol. 488. New York: Farrar, Straus and Giroux.

Humenansky, J. (2022). Learning by daoing: Designing for the unique properties of daos. Medium.

Kreutler, K. (2021). A prehistory of DAOs. Mirror.

Larimer, D. (2013). Larimer, Overpaying for security.

Selkis, R. (2021). Crypto theses for 2022.

Turley, C. (2021). DAO landscape. Mirror.

Chapter 8

The Next Step of Web3 — Metaverse

Jiajian Wang and Qinxu Ding

Metaverse is a new Internet application and social form that integrates various new technologies. It provides an immersive experience based on extended reality technologies, generates a mirror image of the real world based on digital twin technology, and builds an economic system based on blockchain technology.

Metaverse closely integrates the virtual and real worlds in economic, social, and identity systems. It allows each user to produce content and edit the world.

Metaverse, a combination of "meta" and "verse", refers to the "transcendent universe" or "self-evolving universe" (Stephenson, 2003). From this definition, Metaverse entrusts human yearning and cognition to the new universe.

Therefore, it should not be a replica of the real world but a new and viable ecosystem.

It must be emphasized that the metaverse is not equal to the game or the virtual world. Moreover, it is still an evolving concept, with different players constantly enriching its meaning.

How to build a metaverse with the same "life force" as the real universe? At present, the construction of the metaverse is divided into two camps — the classical metaverse camp and the crypto metaverse camp. The former focuses on the exploration to make the metaverse "operate" and tries to solve the productivity contradiction of the metaverse. In contrast, the latter focuses on building the metaverse's economic system and solving the metaverse's production relations.

We believe that these two camps will merge and form a true metaverse in the future.

1.   What Does the Metaverse Look Like?

The metaverse is still relatively abstract. Are there some features and criteria to describe the metaverse? By collating a large amount of public information (Chang & Liu, 2022; Grider & MAXIMO, 2021; Wang & Xiang, 2021; Zhu *et al.*, 2022), we abstract the metaverse into six essential characteristics with broad consensus:

Mapping reality: That is the migration from the real world to the virtual world. The current stage involves avatars, social interaction, collaboration and creation, digital twins, and virtual assets, which are closely related to reality mapping units.

Sensory immersion: Sensory immersion can be divided into three levels. The first is the visual, physical immersion brought about by VR equipment. The second is to build social relationships in the virtual world, resulting in long-term residence and social-based spiritual immersion. Then people start being unable to leave the metaverse and don't care if they need to jump from the metaverse to the real world. At this time, immersion in thought is formed. Finally, there is the ultimate immersion. People can gain the ability to live and thrive in the metaverse to achieve life immersion.

Self-evolution: Self-evolution has three meanings: permanence, self-referentiality, and evolution. There are no Internet states such as "downtime", "interruption", "experience of censorship", or "denial of service" in the metaverse, just like the universe will not stop. The metaverse has only one state — a perpetual state, which will be guaranteed by the underlying architecture built by the blockchain. At the same time, it will continue to evolve like the real universe.

Decentralization of power: In the same vein as self-evolution, decentralization of power refers to establishing the autonomous control of digital rights such as data, assets, identities, and access to services by "citizens" of the metaverse through blockchain technology. It does not need to be controlled by centralized powers.

Open economic system: Many virtual worlds in games have economic systems. The game creators have also been quantified and distributed so that their labor income can be circulated and realized in a relatively closed-loop economic system. However, the economic system of the metaverse should be open.

Borderless interoperability: The borderless eliminates the boundary between virtual and reality, the boundary of the metaverse, the boundary between open and closed, and the boundary of digital assets. Therefore, it establishes interoperability. The current metaverse is clearly fragmented, but the metaverse will eventually be unified just as the Internet moves from the local area network to the Internet. This borderless interoperability will bring the immeasurable marketplace to the metaverse and profoundly change people's traditional consumption habits and lifestyles.

From the above definition, it's clear that we cannot equate the metaverse with the virtual world, as the virtual world describes only part of the metaverse.

Moreover, our understanding of metaverse fully affirms the value of blockchain and crypto culture in the construction of metaverse, and even the blockchain plays a central role in the construction of metaverse. It is the primary factor in the metaverse being called the universe.

2. Classical Metaverse: Big Internet Companies' Imagination of Metaverse

In this chapter, we distinguish between Internet companies and crypto companies in building metaverse and describe them as "classical metaverse camp" and "crypto metaverse camp". This difference is the key to understanding what metaverse is in different contexts.

Looking further into the future, the two camps are not superior or inferior but are exploring the metaverse from different angles.

We believe these two camps will merge and work together to form a true metaverse in the future.

The classical metaverse is led by Internet companies and is derived from the Internet and Web2. Therefore, the "founder" of the classical metaverse also tries to describe the metaverse as an upgraded version of the Internet.

2.1. *The six technical supports of the classical metaverse*

It can be observed that the classical metaverse camp is constantly making efforts on technical infrastructure such as cloud computing, 5G, Internet of Things, artificial intelligence, rendering engine, etc. (Chang & Liu, 2022). These technological advances are becoming crucial technical support for the metaverse. This is also their core contribution to the construction of the metaverse so far.

- **Virtual reality technology**: Virtual reality technology is the entrance to the metaverse and one of the front-end technologies for users.

Virtual reality technology can be divided into three types with differences, namely VR (virtual reality), AR (augmented reality, superimposing virtual things on the real world), and MR (mixed reality, superimposing virtual things on the virtual world). These three are sometimes referred to collectively as XR (extended reality). Integrating the three visual interaction technologies gives users a sense of immersion in the seamless transition between the virtual and real worlds. Currently, XR is the most technical means that can bring users a combination of real and virtual experiences.

- **Artificial intelligence**: Artificial intelligence is one of the underlying technologies of the metaverse.

Artificial intelligence attempts to understand the essence of intelligence, simulate the information process of human consciousness and thinking, and produce a new intelligence machine that can respond similarly to human intelligence, even surpassing human intelligence. Research in this area includes machine learning, natural language recognition, image recognition, natural language processing, etc.

- **5G network**: For the metaverse, the emergence of 5G will make the high-speed interconnection of everything a reality. 5G network has three major application scenarios: high speed, low latency, and wide connection. It is the core technology for metaverse experience improvement.

- **Computing platform**: Computing power is one of the most critical physical origins of the metaverse. The metaverse's demand for computing power will be stronger than any previous technological change. Finding a general computing architecture that can carry super computing power will become an essential mission for the builders of the metaverse's underlying information system.

- **Engine technology**: The game can simulate, extend, and imaginary reality into a three-dimensional and gorgeous virtual world. The most significant contributor behind this is graphics-based engine technology. Engine technology is also essential to support the reality and immersion of the metaverse.

- **Digital twin**: The digital twin is designed to reproduce the real world in the virtual world at a ratio of 1:1 by digitally modeling real industrial production, using the Internet of Things technology to collect massive data, and using large-scale data analysis and artificial intelligence technology. It requires implementations to map the real world in the metaverse. The initial stage of the metaverse is nothing, and the digital twin will provide rich content for the metaverse.

If we look at the technical exploration of the classical metaverse camp in building the metaverse, we can observe that their efforts are mainly focused on making the metaverse work, and the experience is immersive. These technologies are the core and key to the metaverse's future proliferation (see Table 1).

Suppose we divide the metaverse into three layers. In that case, most of today's "work", whether it's a scene, software, or hardware, belongs to the Internet, that is, the category of the traditional metaverse camp.

However, it is clear that these technological explorations are time-consuming and labor-intensive and may not be effective in the short term. This is also one of the reasons why the current market generally

Table 1.   Metaverse technologies.

Application scenarios	Social, work, game, industry, urban governance...		
Software technology	Artificial intelligence	Cloud service	Digital twin
Hardware technology	Perceptual interaction	Network transmission	Chip computing power

lacks recognition of the "metaverse" because these technologies can still not support and present the six characteristics that an ideal metaverse should have.

It has to be mentioned that the classical metaverse has also worked hard on blockchain technology. Still, given that Internet practitioners are often based on alliance chains, we believe this is contrary to the requirements of the metaverse to be open, decentralized, and borderless. So it is not included in the exploration of production relations.

In contrast, most Internet companies in the classical metaverse camp directly face users and have significant influence. They currently develop user-oriented metaverse products with the help of software and hardware technologies such as AR, VR, MR, etc. These products constitute the public's perception of the metaverse.

In short, the classical metaverse took the lead in triggering the productivity revolution of the metaverse through all-around technological innovation. And through the immersive experience, it is the first to let internet users feel the novel experience of the metaverse.

2.2. A list of typical application cases of classical metaverse

Now, we can look at a few typical cases of the classical metaverse camp.

- **Nvidia**: Nvidia launched the Omniverse virtual work platform, which can create a virtual 3D world that conforms to the physical laws of the real world and is highly in line with the real world.

It is one of the most powerful tools for practitioners to construct the virtual world of the metaverse.

- **Meta:** Meta is the banner of the classical metaverse, with the VR terminal Oculus at the hardware level and the metaverse social application at the software level — Horizon Worlds. Oculus users can create an avatar to navigate the virtual world of Horizon or interact with other users' avatars. In February 2022, it was officially announced that the number of monthly active users reached 300,000 for the first time (Andrew, 2022).

- **Tencent:** Through the combination of Super QQ Show, Unreal Engine, and QQ Channel, Tencent QQ users can generate their avatars and then interact with other users in the small scene virtual space built by Unreal Engine. Tencent's metaverse strategy basically continues its advantages in the social field.

- **NetEase:** NetEase recently promoted the immersive conference system "Yaotai". In Yaotai, each participant has his avatar. Users can browse through virtual scenes such as conference rooms and exhibition halls. They can also use text or voice to communicate with other users. NetEase's metaverse play style is also strongly related to its "space creation" ability in the game field.

3. Crypto Metaverse: The "Metaverse Attempt" of the Web3 Pioneers

The crypto metaverse is led by blockchain companies, mainly relying on technologies such as blockchain, NFT, and zero-knowledge proofs. It tries to start from decentralization and economic systems, leading the revolution of production relations in the metaverse.

The achievements of Internet companies are to enable the metaverse to "operate", while blockchain companies try to build the operating rules and economic systems of the metaverse. In contrast, Internet companies are more in line with the current reality, while blockchain companies are highly advanced in concept and technology exploration.

The builders of the crypto metaverse are also the pioneers of Web3. In the previous chapters of this report, we have detailed the outstanding achievements of the crypto metaverse camp. The following is a brief description of the technical value of these Web3s.

3.1. *Six technological contributions of the crypto metaverse*

Blockchain underlying technology: The blockchain's underlying technology is the crypto metaverse infrastructure, such as the public chain Ethereum, Arbitrum in Layer2, decentralized cloud's Filecoin and Arweave, etc. These technical facilities lay the "technical foundation" for the crypto metaverse.

DeFi: The current DeFi mainly involves several protocols such as trading, derivatives, lending, synthetic assets, and stablecoins. They combine to form DeFi Lego, which constitutes the "economic system" of the crypto metaverse, an open financial system.

NFT: NFT technology truly allows users to enjoy unique digital ownership in the Web3 era. It creates native virtual assets such as crypto artwork and brings real-world assets into the metaverse through mapping. NFT is building the asset system of the metaverse.

GameFi: GameFi is creating a more immersive crypto game metaverse, which is keeping the same pace as the practice of the classical metaverse. It is also the most popular subdivision in the current crypto industry. The exciting thing about GameFi is that it tries to give traditional games an open "economic system", which makes the game break away from its own environment. GameFi is a transition version from the crypto metaverse to the real metaverse.

Virtual world: In 2022, virtual world crypto projects are gradually increasing. Decentraland, which issued coins in September 2017, is the most famous among them. It is also the first fully decentralized and user-owned virtual world. This platform allows users to browse, create content, and interact with other people and entities.

DAO: DAO is a new collaborative paradigm created by the pioneers of the crypto metaverse. In the process of building the metaverse, DAO is the best collaborative organization method and provides a detailed and feasible technical implementation. It is very suitable for mobilizing large-scale cross-domain metaverse content creation.

3.2. State of the crypto metaverse: Very early but fast

From a Web3 perspective, where is the metaverse now? Perhaps we can observe the current progress from the number of users. Table 2 shows the number of unique addresses we have sorted out for mainstream Web3 applications.

The above is only the number of addresses on the chain, which does not represent the actual number of unique users. It is not difficult to find that the crypto metaverse is still in a very early stage.

Figure 1 shows the number of addresses of mainstream DeFi applications. It can be found that the number of users of other applications other than Uniswap is not high. However, from the figure, it can also be seen that the number of unique addresses has increased very rapidly since 2020. So it can be observed that the crypto metaverse is developing very fast.

We can also observe the pace of development of the crypto metaverse from another aspect. According to Pitchbook data (Da, 2021), global venture capital investment in crypto projects totaled $10 billion in the first quarter of 2022. This is the largest quarterly investment on record and more than double the level of

Table 2. Number of unique addresses for mainstream Web3 applications.

Project/Category	Number of unique addresses	Remark
Ethereum | Public chain infrastructure	208 million	
Uniswap | DeFi application	4 million	
1INCH | DeFi application	1.34 million	
0x API | DeFi application	0.57 million	
Axie | Game application	0.22 million	
StepN | Game application	0.74 million	Statistics only Solana
OpenSea | NFT application	1.85 million	Statistics only Ethereum
Decentraland | Virtual world	0.8 million	

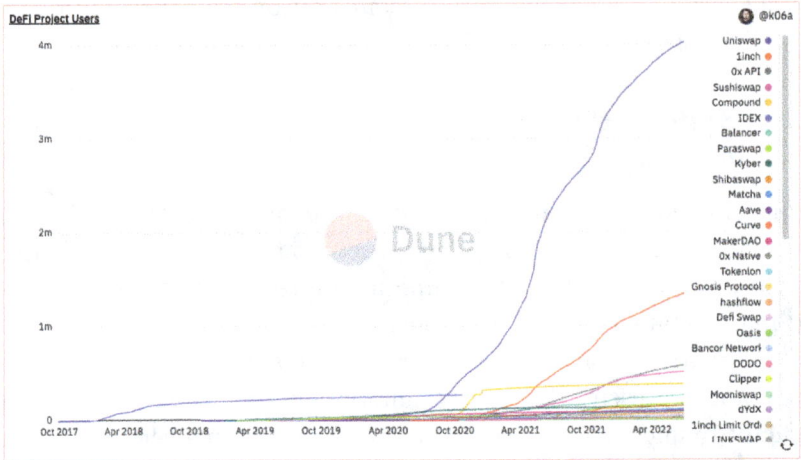

Figure 1.   The DeFi project users of mainstream DeFi applications (https://dune.com/k06a/DeFi-Project-Users).

the same period in 2021. The full-year totals for 2019, 2020, and 2021 were $3.7 billion, $5.5 billion, and $28 billion, respectively.

4.   The Next Step for Metaverse Web3

As mentioned earlier, the result of Internet companies is to allow the metaverse to "operate". They conceived the "metaverse" from the perspective of "interaction" and "experience". On the other hand, blockchain companies or Web3 entrepreneurs are trying to build the metaverse's operating rules and economic systems. It's difficult to say which is better, but the starting point and angle of constructing the metaverse are different. The fusion of the crypto metaverse and the classical metaverse constitutes the complete metaverse that the public expects.

The ultimate metaverse must be a multi-technology fusion, that is, the six characteristics we mentioned above: mapping reality, sensory immersion, self-evolution, decentralization of rights, open economic system, and borderless interoperability.

But it has to be pointed out that based on the technical characteristics of Web3 and the essence of the metaverse, blockchain technology only accounts for a small part of the underlying technology that constitutes the metaverse, but it is undoubtedly the core part.

4.1. Classical and crypto metaverse camps synergize with each other

The classical and crypto metaverse camps currently reach the same goal and integrate and synergize with each other. For example, Twitter has added crypto NFT and crypto payment. GameFi and Decentraland have also tried to bring the decentralization concept and economic model of Web3 and even DAO into more immersive and low-latency applications.

- Web3 practice of well-known Internet companies

Twitter: In September 2021, the BTC tipping function based on the Lightning Network was launched (Zack, 2021). In January 2022, the official verification function of the NFT avatar was launched (Will, 2022).

Instagram: In May 2022, the NFT display function was tested (Aisha, 2022).

- A typical case of Web3 entrepreneurs "replicating" Internet applications

Virtual space — Decentraland: Decentraland can be translated as "decentralized continent" and is a virtual reality platform powered by the Ethereum blockchain. In this virtual world, users can buy and sell land, build facilities, and profit from it. At the same time, users can log in to Decentraland to visit and experience the virtual world inside. Some media commented that it has the potential to become the next big social network, a thriving business city, and the gateway to a new "spatial" Internet (see Table 3).

SocialFi (Decentralized social) — Lens Protocol: The Lens protocol is a user-owned open social graph that can be plugged into any application. On May 18, 2022, the Lens protocol was officially launched on the Polygon mainnet (Vismaya, 2022). In short, it's a blockchain version of Twitter.

Table 3. List of other virtual world projects in the crypto metaverse.

Category	Name
Virtual world/Social	The Sandbox, Decentraland, Cryptovoxels
NFT land	BAYC, Worldwide Webb
Game	Illuvium, My Neighbor Alice, Alien Worlds

SocialFi (Decentralized social) — SecondLive: SecondLive is a 3D virtual space-based social application. It allows users to create avatars in the metaverse and enable multiple virtual spaces, including experiences like games, concerts, seminars, etc. Moreover, it has an NFT marketplace where anyone can buy, sell, and trade digital assets.

4.2. Three stages of metaverse development

So what is the evolutionary path of the real metaverse? It is pointed out that the evolution of the metaverse is the digitization process in human civilization, and the process of human digital migration can be summarized into three stages: digital twin, digital native, and digital immortality (Chang & Liu, 2022).

Therefore, the development of the metaverse will have three corresponding stages.

- **Digital twin**: The digital twin is mapping the real world to the digital world, creating a virtual body of a physical entity. Since the birth of computers, digital twins have never stopped.

However, starting from Web3, digital twins are no longer simple digitization but the mapping of basic substances, assets, and cultural symbols in the real world to the chain. One of the core areas of Web3's digital twin is stablecoins, especially Tether and MakerDAO.

Tether is the asset reserve of the real world, corresponding to the issuance of USDT in the blockchain world, which maps the assets of the real world to the digital world. On May 19, Tether stated that 86% of its reserve assets are cash or cash equivalents, and the rest includes corporate bonds, secured loans, etc. (Stacy, 2022) (partly investment in assets such as virtual currencies). Although

MakerDAO is an algorithmic stablecoin, it has also added real-world asset reserves during its development.

NFTs are another booming asset mapping system. For example, the traditional museum's NFTization of famous paintings is to map the real culture to the digital world.

- **Digital native**: Digital native is the creation of new knowledge beyond reality and human cognition. For Web3, this process is to create art, assets, cultural IP, and business models that do not exist in reality. The metaverse is not about replicating an actual universe but about creating a universe that doesn't exist currently.

Digital native is prominent in the Internet world, but the Web3 field has created many digital assets. We believe that DeFi, GameFi, SocialFi, and NFT are all critical components of the digital native. But we are not going to give further expansion in this report.

- **Digital immortality**: Digital immortality means that human life is no longer marked by physical life. With the development of science and technology, human consciousness and memory can also be uploaded into the metaverse. Finally, the fusion of the digital world and the real world will be realized, and humans can gain digital immortality in the metaverse. On the other hand, virtual characters will also have autonomous consciousness. They will seek to jump from the virtual world to the real world, and digital life may also have the same rights as real life.

American TV series "Upload" tells the story of when human beings are going to die, they can "upload" all memory and consciousness into digital space. Digital space is another human society. They can also interact with relatives and friends in real space in a visual scene at any time, thereby realizing digital immortality.

Digital immortality forms the true metaverse, which will be the end result of Web3.

We believe the metaverse development has no apparent timeline relationship in the above three stages. Still, digital immortality will result from digital twins and digital natives developing to a particular stage. However, digital twins and digital natives will continue to evolve after digital immortality is realized.

References

Aisha, M. (2022). Instagram to start testing NFTs with select creators this week. https://techcrunch.com/2022/05/09/instagram-testing-nfts-select-creators/.

Andrew, H. (2022). Meta's VR worlds are growing fast, with usage climbing above 300k monthly actives. https://www.socialmediatoday.com/news/metas-vr-worlds-are-growing-fast-with-usage-climbing-above-300k-monthly-a/619145/.

Chang, J. & Liu, Q. (2022). Yuan Yu Zhou, Tong Wang Wu Xian You Xi Zhi Lu [Metaverse, the Road to Infinite Gaming]. Zhong Xin Chu Ban Ji Tuan. (in Chinese).

Da, K. (2021). Report: In 2022, crypto investment will become mainstream, and the total amount of financing in the crypto market in 2021 will exceed the sum of previous years! (in Chinese). https://view.inews.qq.com/a/20211228A05AGD00.

Grider, D. & Maximo, M. (2021). The metaverse: Web 3.0 virtual cloud economies. Grayscale Research.

Stacy, E. (2022). New Tether report shows 17% reduction in commercial paper. Retrieved from https://decrypt.co/100869/new-tether-report-shows-17-reduction-commercial-paper.

Stephenson, N. (2003). *Snow Crash: A Novel*. Spectra.

Vismaya, V. (2022). Aave's decentralized lens protocol goes live on polygon. https://www.cryptotimes.io/aaves-decentralized-lens-protocol-goes-live-on-polygon/.

Wang, R. & Xiang, A. (2021). Yuan Yu Zhou Fa Zhan Yan Jiu Bao Gao [2021 Metaverse Development Research Report]. Qing Hua Da Xue Xin Mei Ti Yan Jiu Zhong Xin. (in Chinese).

Will, G. (2022). Twitter launches NFT profile picture verification. https://www.coindesk.com/business/2022/01/20/twitter-launches-nft-profile-picture-verification/.

Zack, S. (2021). Twitter to add bitcoin lightning tips, NFT authentication. Retrieved from https://www.coindesk.com/tech/2021/09/23/twitter-to-add-bitcoin-lightning-tips-nft-authentication/.

Zhu, J., Lu, K., Gao, R., & Kang, Y. (2022). Chu Tan Yuan Yu Zhou [Exploring the Metaverse]. KPMG China. (in Chinese).

Chapter 9

Industry Regulation

Richard Li and Brian Heng

Various types of financial businesses and services developed around cryptocurrencies provide the financial system necessary for the development of the Web3 World.

At present, it seems that only El Salvador and the Central African Republic recognize Bitcoin as legal tender. Even though many countries still do not recognize Bitcoin or other forms of cryptocurrency and award them the fiat currency status, they do not restrict the uses or trading of cryptocurrencies. In many instances, cryptocurrencies are also being considered as a legitimate asset.

However, the problems arising from virtual finance, including speculative risks, money laundering and other financial crimes, are increasingly catching the attention of regulatory authorities to monitor its downsides.

While at the same time, the regulatory authorities in various countries are promoting the creation of Web3, as an emerging field in innovative asset developments, in order to grasp the immense opportunities and reap the benefits riding on the next wave of Internet revolution. The key challenge for the authority is to find a balance between mitigating risks and encouraging innovation in the field.

Due to space limitations, this report mainly focuses on discussing some of the world's major economies' key regulatory requirements and attitudes.

1. United States of America

In December 2021, the United States of America (US) House of Representatives held a crypto asset hearing. Republican Representative Patrick McHenry in his opening remarks, said that "2021 is the year of cryptocurrency" and that "of course, we need reasonable rules of the road, but the knee-jerk reaction of lawmakers to regulate out of fear of the unknown will only stifle American ingenuity and put us at a competitive disadvantage". He raised a core concern to the interest of United States, "How do we make sure the cryptocurrency revolution happens in the United States and not overseas?".[1]

Overall, the US attitude toward cryptocurrencies and crypto finance is to mitigate risks on one hand, while on the other hand to provide guidance for the innovation and developments of the industry.

In the United States, cryptocurrency trading is not illegal. The United States is a country comprises of a federation of states. The regulatory attitude of cryptocurrency finance varies across states. Though at the federal level the laws and policy developments of cryptocurrencies are constantly evolving and advancing.

At the federal level, different regulatory authorities characterize cryptocurrencies and their transactions differently. For example, the US Securities and Exchange Commission (SEC) characterizes cryptocurrencies as securities, so digital wallets and exchanges are fully subject to securities laws. Compared with the US Commodity Futures Commission (CFTC), which describes Bitcoin as a commodity and allows cryptocurrencies to be offered as derivatives to the public for trading.

---

[1]McHenry at Digital Assets Hearing: 2021 Was the Year of the Cryptocurrency. https://republicansfinancialservices.house.gov/news/documentsingle.aspx?DocumentID=408215.

For the Financial Crimes Enforcement Agency (FinCEN), a division of the US Department of the Treasury, cryptocurrency exchanges are treated similarly with traditional currencies under the same supervision mechanism, and all are subject to the provisions of the Bank Secrecy Act, which the transaction service providers must comply with anti-money laundering, counter-terrorism financing, etc. According to the regulations, the providers are required to collect and share relevant cryptocurrency transactions, including the information of the initiators and beneficiaries, with the authority.

The US Treasury Department has clearly emphasized the urgent need to promote the development of regulations for crypto finance in future. This is to respond to criminal activities at home and abroad. The federal government is focusing on the revolutionary innovations that may bring about on the economic developments by crypto finance and Web3. Thus, the United States is in favor of a unified supervision instead of the current fragmented regulatory supervision by different authorities.

In March 2022, the US President Joe Biden signed the Executive Order on "Ensuring the Responsible Development of Digital Assets", consisting of six key broad policy objectives: protecting consumer and investor rights; enabling financial stability; combating financial crimes; ensuring US financial leadership and competitive advantage; promoting responsible innovation; and supporting financial inclusivity.[2]

In response to the Executive Order, the US Department of Commerce's Center for International Trade Management issued Developments in May announced the "Developing a Framework on Competitiveness of Digital Asset Technologies"[3] and solicited for comments on 17 questions addressing the competitiveness of crypto

---

[2]Executive Office of the President: Ensuring responsible development of digital assets. https://www.federalregister.gov/documents/2022/03/14/2022-05471/ensuring-responsible-development-of-digital-assets.

[3]International Trade Administration, Department of Commerce: Developing a framework on competitiveness of digital asset technologies. https://www.federalregister.gov/documents/2022/05/19/2022-10731/developing-a-framework-on-competitiveness-of-digital-asset-technologies.

companies in the United States, and many other issues, such as the advantages and disadvantages of crypto assets as compared to traditional finance, and their respective technological progresses.

On June 7, 2022, Republican Senator Cynthia Lummis and Democratic Senator Kirsten Gillibrand co-sponsored the "Responsible Financial Innovation Act".[4] As the first major bipartisan legislative proposal, the bill has received widespread attention and, if passed, will be the formal laws governing crypto assets. For crypto assets defining as "commodities", the Commodity Futures Trading Commission (CFTC) will be granted exclusive jurisdiction.

To address the regulatory overlap between the CFTC and SEC, the new bill introduces "ancillary asset" concept. "Digital asset" that deems as ancillary asset will be under the CFTC jurisdiction, however, the issuer of those ancillary assets is required to disclose legally required information to the SEC twice a year.

In addition, under the Lummis-Gillibrand Act,[5] "payment stablecoins" will also be subject to more prudent regulation. For example, depository institutions that issue stablecoins must maintain high-quality liquid assets of not less than 100% of the par value of the issuance, and monthly disclose stablecoin-backed assets and their values, and the number of unredeemed stablecoins, etc.

It is worth noting that the regulation of stablecoins is also a concern of the Biden's administration at present. The US President's Working Group on Financial Markets in its November 2021 report, cited US Treasury Secretary Yellen's view on stablecoins, said "the absence of appropriate oversight presents risks to users and the broader system," and "current oversight is inconsistent and fragmented, with some stablecoins effectively falling outside

---

[4]Lummis-Gillibrand Responsible Financial Innovation Act. https://www.congress.gov/bill/117th-congress/senate-bill/4356/text?r=1&s=1#toc-idce30b4d250cc4912b51c0b1ff159e0eb.

[5]The Lummis-Gillibrand Responsible Financial Innovation Act: What to know. https://www.mayerbrown.com/en/perspectives-events/publications/2022/06/the-lummisgillibrand-responsible-financial-innovation-act-what-to-know.

the regulatory perimeter".[6] In response to the potential risks of stablecoins, the report proposes some specific legislative recommendations and interim measures.

As the head of the US central banking system, Fed Chairman Jerome Powell seems to be more cautious about cryptocurrencies. At a congressional hearing in June 2022, Powell said he did not believe price volatility in Bitcoin and other cryptocurrencies has had a macroeconomic impact, but it is believed that there must be a better regulatory framework for cryptocurrencies. For stablecoins, Powell said this is a new thing, but there are no corresponding regulatory measures in place.

Powell also believed that the central bank digital currency (CBDC) is what the United States should really explore regardless of partisanship. Federal Reserve plans to conduct research on relevant policy and technical issues in the coming years and submit recommendations to Congress in due course. He felt that the digital dollar should be guaranteed by the government, not of those stablecoins that issued by private institutions ending up as digital dollars.

At the Bank for International Settlements Innovation Summit in March 2022, Powell expressed that the Fed has always supported responsible innovation, but the existing regulatory framework was not established with the digital world in mind, and new rules, laws, and frameworks must be put in place to cope with the risks arising from stablecoins and other cryptocurrencies that may pose for the US financial system and investors.

2. European Union

Regulatory policies on cryptocurrencies vary from member states in the European Union (EU), but in general, cryptocurrencies are seen as a legal asset. In some Euro countries, cryptocurrency exchanges need to register with the financial regulatory authorities of the

---

[6]President's Working Group on Financial Markets Releases Report and Recommendations on Stablecoins. https://home.treasury.gov/news/press-releases/jy0454.

host country. In the EU, cryptocurrencies and crypto assets are considered Qualified Financial Instruments (QFI). EU laws do not prohibit banks, trusts, and investment companies to hold crypto assets, cryptocurrencies, or to provide related services.

However, operators engaging in cryptocurrencies related business must comply with a series of EU's regulatory requirements. The issue of financial crimes involving cryptocurrencies has always been a key focus in the EU.

The European Union's Anti-Money Laundering Directive 5 (5AMLD), which came into effect in January 2020, categorizes the exchange and trading of cryptocurrencies and fiat currencies to be under regulatory supervision. Exchanges are required to carry out due diligence work such as KYC/CDD on customers. The Anti-Money Laundering Directive No. 6 (6AMLD), which came into force in December of the same year, further strengthens the compliance regulations.

In July 2021, the European Commission aimed at strengthening the EU's anti-money laundering and counter-terrorism financing supervision includes the cryptocurrency industry. The four legislative proposals include the establishment of a core EU-level anti-money laundering regulatory unit, as well as customer due diligence, new anti-money laundering directives, and setting caps on large payments and cash movements, etc.

The EU tried to introduce a unified regulatory framework in response to the problems of fragmented supervision within the EU.

The European Commission introduced a proposal in September 2020 known as the Markets in Crypto-Assets Regulation (MiCA) proposal. This is the most comprehensive regulatory framework for crypto assets at EU level.

On June 30, 2022, the President of the European Council and the European Parliament reached a provisional agreement on the MiCA[7] proposal, which for the first-time placed crypto assets, crypto

---

[7]Digital finance: Agreement reached on European crypto-assets regulation (MiCA). https://www.consilium.europa.eu/en/press/press-releases/2022/06/30/digital-finance-agreement-reached-on-european-crypto-assets-regulation-mica.

asset issuers, and crypto asset service providers under the same regulatory framework. The proposal covered unsecured crypto assets, stablecoins, and crypto trading, and wallets where crypto assets are held. The Council of the European Union in its press release states that new regulatory framework will protect investors and maintain financial stability, while allowing innovation and raising attractiveness of the crypto industry.

In addition, although the EU does not currently have a total ban on Proof of Work (PoW) in the sector, those mining activities that consume a lot of electricity, and are environmental unfriendly do not seem to be popular with the EU regulators.

In February 2022, EU Interior Commissioner Ylva Johansson told the Munich Security Conference[8] that the EU was open to cryptocurrencies, including Bitcoin, but only if the crypto financial activities are properly regulated and supervised in order not to give criminal organizations or terrorist groups an opportunity to take advantage of the anonymity.

However, it is worth noting that ECB President Christine Lagarde in May 2022 in an interview said that cryptocurrencies have "no value"[9] and that there are no underlying assets to support the security of the crypto currency. She called on global policymakers to come up with rules to protect inexperienced investors.

Her attitude also represents the ECB's overall approach to crypto assets. The European Central Bank (ECB) announced in May 2022 through "The Financial Stability Review" states that crypto assets lack intrinsic economic value and have been frequently use as a speculative tool, resulting in price fluctuations, high energy

---

[8]Commissioner Johansson's opening remarks at the Munich Security Conference 2022 panel conversation, "Gain or Pain": Security-Proofing National Cryptocurrencies. https://ec.europa.eu/commission/commissioners/2019-2024/johansson/announcements/commissioner-johanssons-opening-remarks-munich-security-conference-2022-panel-conversation-gain-or_en.

[9]Christine Lagarde says crypto is worth nothing. https://www.cnbc.com/2022/05/23/ecb-chief-christine-lagarde-crypto-is-worth-nothing.html.

consumption and illegal financial activities; and, thus resulting the high-risk nature of crypto assets.[10]

The Review argued that TerraUSD's recent collapse shows that stablecoins are not really stable in long-term. If the size and complexity of the crypto-asset markets continue to grow, and financial institutions are increasing interconnected with crypto asset markets, this will pose a systemic risk to the financial stability.

3. United Kingdom

After Brexit, the United Kingdom (UK) converted the regulatory requirements on cryptocurrencies in the EU Anti-Money Laundering Directives 5 and 6 as laws to serve its own country's needs. Therefore, crypto asset companies setting up in the UK or providing services to the UK clients will require to register with the Financial Conduct Authority (FCA).

They will need to comply with anti-money laundering and counter-terrorism financing, etc. The FCA only when satisfied with the applicant's established monitoring mechanisms, such as identifying the criminal or terrorist groups, under its anti-money laundering regulations will then grant registration to the applicants.

In January 2022, the UK government announced that it would strengthen regulatory requirements for crypto asset advertising, which will have to observe to the same standards as stocks, insurance, and other financial products to protect consumers from being misled through publicity.[11] This is to ensure that the interests of consumers are protected without stifling innovation. Currently, the FCA is soliciting feedback on its new crypto publicity regulations. Under the

---

[10]Decrypting financial stability risks in crypto-asset markets. https://www.ecb.europa.eu/pub/financial-stability/fsr/special/html/ecb.fsrart202205_02~1cc6b11 1b4.en.html.

[11]Cryptocurrency regulation UK. https://investingreviews.co.uk/guides/cryp tocurrency-regulation-uk/; Government to strengthen rules on misleading cryptocurrency adverts. https://www.gov.uk/government/news/government-to-strengthen-rules-on-misleading-cryptocurrency-adverts.

new rules, both cryptocurrency exchanges and providers of crypto-asset services must ensure that customers are clearly aware of their risk exposures.

In April 2022, the UK government announced on its website that the UK Treasury will take a series of measures to make the "UK a global hub for crypto-asset technology and investment".[12] These measures include bringing stablecoins into regulation, eventually making them recognized as a form of payment method. Others include the introduction of financial market infrastructure sandbox for firms to experiment with innovation; the establishment of a crypto-asset engagement group to work with industry; and, the launch of Non-Fungible Token (NFT) in partnership with Royal Mint.

4. China

To guard against any potential financial risks, China has shown a near-negative regulatory stance toward cryptocurrency-related financial activities. In 2013, People's Bank of China and five other governmental agencies issued the "Notice on Preventing Bitcoin Risks", clarifying that Bitcoin should be a specific virtual commodity, does not have the legal status equivalent to fiat money, and cannot and should not be used as money circulating in the market. The notice clearly requires financial institutions and payment institutions not to buy or sell bitcoins, or provide its related services.[13]

In September 2017, the People's Bank of China and seven other governmental agencies jointly issued the "Announcement on Preventing the Financing Risk of Token Issuance"[14] that confirms

---

[12]Government sets out plan to make UK a global cryptoasset technology hub. https://www.gov.uk/government/news/government-sets-out-plan-to-make-uk-a-global-cryptoasset-technology-hub.

[13]People's Bank of China Issues a regulatory notice on bitcoin. https://thediplomat.com/2013/12/peoples-bank-of-china-issues-a-regulatory-notice-on-bitcoin/.

[14]Briefing on the People's Bank of China notice on further preventing and handling the risk of speculation in virtual currency transactions. http://www.pbc.gov.cn/goutongjiaoliu/113456/113469/4348521/index.html.

token issuance is essentially an act of illegal public financing without approval, including those issued as securities, for fund-raising, and other illegal and criminal activities.

In May 2021, a meeting of the Financial Stability and Development Committee of the State Council announced the crack down on Bitcoin mining and trading activities, resolutely preventing individual risk exposure to Bitcoin from becoming a systemic societal risk. From May to June, across China Bitcoin mining was Re prohibited through policy rectification. In September of the same year, the National Development and Reform Commission also issued a document requiring provincial governments across the country to rectify virtual currency mining activities. Currently virtual currency mining has been largely banned in China.

While China has strict regulation of crypto finance, it has strong support for blockchain which is the underlying technology of cryptocurrencies.

The "14th Five-Year Plan"[15] lists blockchain as one of the seven key industries of the digital economy, which include cloud computing, big data, and Internet of Things, industrial Internet, artificial intelligence, and virtual reality and augmented reality.

Regarding the specific development of blockchain, the 14th Five-Year Plan proposes to promote smart contracts, consensus algorithms, encryption algorithms and distributed fault-tolerance mechanisms as blockchain technology innovations. Through alliance and collaboration to focus on developing blockchain service platforms and finance technologies, supply chain management, government services within the focus of alliance to improve the regulatory mechanism.

5. Hong Kong SAR, China

In November 2018, the Securities and Futures Commission (SFC) of Hong Kong SAR issued the "Statement on regulatory framework

---

[15]China's 14th Five-Year Plan: Blockchain and digital currency part of 'new infrastructure' investments. https://www.blockchaintechnology-news.com/2021/04/chinas-14th-five-year-plan-blockchain-and-digital-currency-part-of-new-infrastructure-investments.

for virtual asset portfolios managers, fund distributors and trading platform operators regulatory"[16] identifying significant risks associated with virtual asset investments. The SFC is issuing guidance on the regulatory standards expected of virtual asset portfolio managers and fund distributors.

The SFC's attitude was to cooperate with those operators, which have the intention and consistently demonstrated their commitment to comply to strict virtual asset trading standards, by incorporating them into the SFC regulatory sandbox".[17] The SFC is also exploring a conceptual framework for the potential regulation of virtual asset trading platform operators.

In November 2020, the Financial Services and the Treasury Bureau (HKBPO) of Hong Kong SAR issued a legislative consultation paper with plans to amend its Hong Kong's Anti-Money Laundering Ordinance and enact to introduce the "Virtual Asset Service Provider Licensing System".[18] In May 2021, the Hong Kong's Treasury Bureau issued a summary of the consultations and will submit the amendment bill to the Legislative Council for consideration during the 2021–2022 legislative year.

The HKSAR Government recommends that cryptocurrency exchanges operating in Hong Kong must obtain permission from Hong Kong market regulators, and services can only be provided to professional investors. For the legislative direction, the HKSAR Government proposes to give the SFC the power to intervene, including restricting or prohibiting the operation of licensed virtual asset service providers when required. But after that, there has been no further significant progress made.

The Hong Kong's Monetary Authority (HKMA) released a discussion paper on crypto assets and stablecoins on January 12,

---

[16]Statement on regulatory framework for virtual asset portfolios managers, fund distributors and trading platform operators. https://www.sfc.hk/en/News-and-announcements/Policy-statements-and-announcements/Statement-on-regulatory-framework-for-virtual-asset-portfolios-managers.

[17]SFC Regulatory Sandbox. https://www.sfc.hk/en/Welcome-to-the-Fintech-Contact-Point/SFC-Regulatory-Sandbox.

[18]Anti-money laundering views sought. https://www.news.gov.hk/eng/2020/11/20201103/20201103_193443_472.html.

2022. The public is invited to comment on the conceptual regulatory framework for stablecoins using for payment purposes.[19]

Subsequently, on January 28, the SFC and the Hong Kong Monetary Authority issued the joint circular on intermediaries' virtual asset-related activities, which provides guidance for intermediaries to engage in virtual asset-related activities in reflection of the regulator's focus.[20] Thus, possible heralding a roadmap for Hong Kong's future development of virtual asset regulation.

The joint circular focuses on three areas: (1) the distribution of virtual asset-related products; (2) provision of virtual asset trading services; and (3) provision of advisory services on virtual assets.

6. South Korea

South Korea tends to adopt a strict regulatory stance toward crypto financial services. For example, in the recent collapse of the Terra ecosystem, the relevant regulatory authorities and legislative departments of South Korea immediately conducted investigation of the incident.

Although crypto exchanges are not banned, they are regulated by the Korea's Financial Supervision Agency (FSA) and operators are required to register with them. In South Korea, crypto exchanges must comply with anti-money laundering and counter-terrorism financing regulations and obtain permission to operate from the Korea's Financial Services Commission.

As early as early 2018, the South Korean government banned cryptocurrency trading from being carried out through anonymous accounts. The real-name authentication system for cryptocurrency accounts was activated to prevent cryptocurrencies from being used for illegal financial activities such as money laundering.

---

[19]Hong Kong Monetary Authority Releases Discussion Paper on Crypto Assets and Stable Coins. https://www.sw-hk.com/zh/20220117-1/.
[20]Joint circular on intermediaries' virtual asset-related activities. https://www.hkma.gov.hk/media/eng/doc/key-information/guidelines-and-circular/2022/20220128e2.pdf.

However, the South Korean government has shown great enthusiasm in the field of metaverse, which is closely related to crypto finance. The government of South Korea in January 2022 unveiled the Metaverse: New Industry Leadership Strategy,[21] aiming to become the fifth largest market in the world by 2026. Under the plan, South Korea will support at least 220 technology companies with sales of more than 5 billion won, and help global meta-universe start-ups entering the Korean market.

7.   Japan

Japan's regulatory attitude toward cryptocurrencies is relatively open. For cryptocurrencies, Japan's relatively legislative amendments focus on the Payment Services Act (PSA) and the Financial Instruments Exchange Act (FIEA) with effective from May 2020.[22] The main amendments centered at the regulation of cryptocurrency custodian service providers and cryptocurrency derivatives activities. Although custodian service providers are not involved in the cryptocurrency trading, but are governed by the PSA, while the derivatives activities are governed by the FIEA.

Japan recognizes Bitcoin and other cryptocurrencies as legal assets and means of payment under the PSA.

Cryptocurrency exchanges are legal in Japan and are regulated by the Financial Services Agency (FSA). Presently, exchange operators required to registered with FSA and will undergo strict qualification assessments, including anti-money laundering, counter-terrorism financing checks, and cybersecurity requirements.

In April 2020, Japan's FSA announced the approval of the establishment of two self-regulatory organizations in the crypto industry — Japan Virtual and Crypto Assets Exchange Association

---

[21]South Korea unveils long-term road map to become world's 5th-biggest metaverse market. https://www.coinspeaker.com/south-korea-metaverse-market/.
[22]Japan implements significant changes to cryptocurrency regulation today. https://news.bitcoin.com/japan-changes-cryptocurrency-regulation/.

(JVCEA) and Japan Security Token Offering Association (STO).[23] The two self-regulatory organizations will work closely with the FSA to implement industry regulatory standards.

On June 3, 2022, the Japanese Senate passed an amendment to the PSA,[24] clarifying the legal positioning of stablecoins, as well as strengthening the regulatory supervision to protect users. According to the latest amendments, Japan allows licensed banks, registered transfer institutions, and trusts as the issuer of stablecoins. Intermediaries responsible for the circulation of stablecoins need to fulfill registration obligations and adhere to stricter anti-money laundering surveillance measures.

8. Singapore

The trading and exchange of cryptocurrencies are legal in Singapore. In 2019, the International Commercial Court of Singapore ruled on cryptocurrencies has the essential characteristics of intangible assets, and that clarified the legitimacy of holding crypto assets.[25]

As one of the global financial hubs, Singapore is open to cryptocurrencies and crypto financing, and is even acknowledged by the industry as the "cryptocurrency paradise".[26] However, in the face of the emerging risks of cryptocurrencies, the government is gradually building more complete and robust regulatory system.

Under the Payment Services Act 2019 (PSA), exchanges and other crypto financial institutions are subject to Monetary Authority of Singapore's (MAS) regulation and must be obtained permit from

---

[23] Japan: FSA Recognizes two crypto self regulatory organizations. https:// crypto.news/japan-fsa-two-crypto-self-regulatory-organizations.

[24] Japan adopts law to regulate stablecoins for investor protection. https:// asia.nikkei.com/Spotlight/Cryptocurrencies/Japan-adopts-law-to-regulate-stable coins-for-investor-protection.

[25] Civil Appeal No. 81 of 2019. https://www.elitigation.sg/gd/sic/2020_SGCAI_2.

[26] "Cryptocurrency Paradise" Singapore was affected by notable crashes. https:// etcshowchoir.com/cryptocurrency-paradise-singapore-was-affected-by-notable-crashes.

them to operate from January 2020 onward. In January 2021, the MAS made amendments to the PSA to extend the coverage of the regulations to cryptocurrency transfers, escrow wallet services, etc. Cryptocurrencies as capital market products, such as initial coin offerings (ICO), is regulated by MAS under the Securities and Futures Act.

MAS issued circulars in 2018 and 2020 to alert the public of the risks of investing in crypto assets. In January 2022, MAS issued guidelines that significantly restricted crypto financial institutions from publicizing their services to the public, for example, the crypto financial institutions are advised not to place advertisements on public transport, websites, social media platforms, radio or print media.[27]

In March 2022, MAS issued Circular PSN02,[28] requiring crypto payment token service providers to comply with anti-money laundering and counter-terrorism financing guidelines, such as providing risk assessment and mitigation mechanisms, customer due diligence, proper records of transactions, and to report on suspicious transactions, etc.

In April 2022, Singapore's legislature made a second reading at the bill[29] on the Financial Services and Markets Act, expanding the regulatory authority of MAS and strengthening the supervision on cryptocurrency service provider, requesting them to provide anti-money laundering/counter-terrorism financing measures, and comply with the strengthened requirements on technical risk management, etc.

---

[27]MAS bans crypto trading platforms from advertising in public areas. https:// fintechnews.sg/58584/blockchain/mas-bans-crypto-trading-platforms-from-adverti sing-in-public-areas.

[28]Notice PSN02 Prevention of Money Laundering and Countering the Financing of Terrorism — Digital Payment Token Service. https://www.mas.gov.sg/regula tion/notices/psn02-aml-cft-notice---digital-payment-token-service.

[29]"Financial Services and Markets Bill" — Second Reading Speech by Mr Alvin Tan. https://www.mas.gov.sg/news/speeches/2022/financial-services-and-mark ets-bill-second-read.

In a reply letter to MPs in July 2022,[30] the MAS stated that it has always believed that cryptocurrencies are not appropriate for general public, and it is carefully considering the introduction of further consumer protection measures, which may include the restriction on retail investors participating in cryptocurrencies trading and regulating the use of those cryptocurrencies for leverage.

---

[30]Reply to Parliamentary Question on restrictions on cryptocurrency trading platforms to protect members of public. https://www.mas.gov.sg/news/parlia mentary-replies/2022/reply-to-parliamentary-question-on-restrictions-on-crypto currency-trading-platforms-to-protect-members-of-public.

Appendix 1

Introduction of Web3's Major Investors

Wendy Wang

In order to materialize the "product first and the token economy second" objectives, the majority of Web3 teams in the early stages of development or just getting started choose to get funds from investment institutions rather than indiscriminately issue tokens. As a result, a number of headline investors have emerged in the course of Web3's development. They tend to cast a wide net and tend to get involved in projects at an early stage, looking for Web3 headliners and stalwarts. The appendix will list some of the top Web3 investment firms and give a brief introduction so that readers can better understand them.

Note: Usually the investment institutions are unique in their projects, but this does not mean that all of their investments will achieve 100% returns, so readers should not follow their portfolios blindly, but only for reference purposes.

The following investment institutions are listed in alphabetical order:

- **a16z crypto**[1]

a16z crypto is the crypto fund of top VC a16z. a16z, known as Andreessen Horowitz, was founded in 2009 by Marc Andreessen and Ben Horowitz in Silicon Valley, and invests in consumer, biomedical, fintech, gaming, crypto, etc. Its various funds currently manage about $33.3 billion in assets (data from official profile).

a16z has launched a total of four crypto funds, the most recent of which was announced in May 2022 with a fund size of $4.5 billion. This brings a16z crypto's assets under management to $7.6 billion.

a16z crypto believes that we are currently in the third phase of Internet development, the Web3 era. This era will unleash a new wave of creativity and entrepreneurship. a16z crypto's portfolio includes well-known Web3 leaders such as Yuga Labs, Optimism, MakerDAO, OpenSea, and Uniswap.

- **Alameda Research**[2]

Alameda Research, which combines investment and trading, is a sister company to the cryptocurrency derivatives exchange FTX, also founded by Sam Bankman-Fried (SBF).

Besides being a ruthless trading machine, Alameda Research also invests in a large number of crypto-related projects, with investments ranging from DeFi, public chains, CeFi, Web3, etc. Its investment layout in the Web3 space includes Immutable X, DAOSquare, Jambo, Mask Network, etc.

- **Animoca Brands**[3]

Animoca Brands was founded in 2014 by founder Yat Siu, a Hong Kong entrepreneur and angel investor. Animoca Brands is one of the

---

[1] Official website: https://a16zcrypto.com/.
[2] Official website: https://www.alameda-research.com/.
[3] Official website: https://linktr.ee/animocabrands.

biggest winners in the NFT and metaverse concept boom of 2021, with The Sandbox being one of the hottest NFT projects at the moment, in addition to being the developer of Axie Infinity.

In addition to project development, Animoca Brands has also become an investor in NFT and metaverse-related projects and is an early investor in star projects such as Sky Mavis, Decentraland, Dapper Labs, and has invested in over 150 NFT and metaverse projects.

• Binance Labs[4]

Like Coinbase Ventures, Binance Labs is based on Binance, the leading exchange in the crypto space. Founded in 2017, Binance Labs has invested in more than 180 projects in 25 countries around the world over the course of its nearly five years of operation.

The Binance Labs portfolio is very diverse and ranges from the most basic DeFi crypto Lego to the latest NFT, gaming, and metaverse, with investments in projects such as Axie Infinity, STEPN, LayerZero, and more.

• Coinbase Ventures[5]

As the investment arm of veteran cryptocurrency exchange Coinbase, Coinbase Ventures has been in the spotlight with each of its strikes. Founded in 2018, the big difference between Coinbase Ventures and other investment firms is that Coinbase Ventures doesn't look like a company: as it has no fixed fund size and zero full-time employees, which is inextricably linked to the decentralized philosophy it supports.

According to Coinbase, Coinbase Ventures closed over 70 investments in the first quarter of 2022 alone and has made over 300 investments by September 2022.

Coinbase Ventures' portfolio includes OpenSea, Starkware, Dapper Labs, Immutable, and more.

---

[4]Official website: https://labs.binance.com/.
[5]Official website: https://www.coinbase.com/ventures.

- ### DeFiance Capital[6]

DeFiance Capital is one of the venture funds active in Web3 and crypto investments, with a focus on DeFi and Web3 games. DeFiance Capital believes that as software has engulfed the world in the last decade, DeFi will engulf traditional finance in the next decade.

In addition, DeFiance Capital believes that Web3 games are changing the gaming industry by redefining the value proposition for players and developers.

DeFiance Capital's investments include Axie Infinity, Mintable, Ultiverse, Balancer, and more.

- ### Delphi Digital[7]

Delphi Digital is a typical research-based investment firm that specializes in investing by conducting in-depth research on various tracks. They have analysts who specialize in all areas of digital assets and can break down and evaluate a project, from specific use cases and economic models to their community and cultural impact, nothing escapes their scrutiny.

Delphi Digital's portfolio includes Immutable X, Axie Infinity, YGG, Illuvium, and more.

- ### Digital Currency Group[8]

Digital Currency Group (DCG) was founded in 2015 by digital currency pioneer Barry Silbert. DCG's current main businesses are CoinDesk (industry-leading media, research and events platform), Genesis Trading (industry-leading institutional lending and brokerage firm), Grayscale (the largest digital currency asset manager), Foundry (a financing and advisory firm focused on digital asset mining and collateralization), and Luno (the world's leading digital asset CEX and wallet).

---

[6]Official website: https://www.defiance.capital/.
[7]Official website: https://delphidigital.io/ventures.
[8]Official website: https://dcg.co/.

Barry Silbert hopes to establish DCG as the Berkshire Hathaway of the cryptocurrency space. DCG has also been the most active investor in the digital currency industry, investing in over 150 companies in 30 different countries. dCG's investments in the Web3 space are focused on the gaming sector, including Dencentraland, Horizon Blockchain Games, etc.

- **Dragonfly Capital**[9]

Dragonfly Capital has concentrated on a different industry with each tranche from the first $100 million fundraising in October 2018 to the second $225 million fund in 2021 to the new $650 million fund in April 2022. Currently, they are focusing on Web3 and the metaverse.

Dragonfly Capital's team is global in nature, with two managing partners: Bo Feng, a legend in China's early Internet era, and Haseeb Qureshi, a professional poker player-turned-investor. The team's motto is "global from day one" since it is crucial to think globally from the start if you want to succeed in the cryptocurrency industry.

Dragonfly Capital's portfolio includes Compound, dYdX, 1inch, Matter Labs, and others.

- **Framework Ventures**[10]

Just two years old, Framework Ventures has already started raising its third fund in April 2022 with an estimated size of $400 million, $200 million of which will be used in the blockchain gaming space. This will also bring the size of its fund to $1.4 billion.

Michael Anderson, co-founder of Framework Ventures, said he believes the next phase of the blockchain industry will attract many new users, and that blockchain gaming is still early and full of opportunity. The economics model of P2E blended with fun 3A games will bring explosive growth to the space.

The company's investments in the Web3 space include RabbitHole, Illuvium, and others.

---

[9]Official website: https://www.dcp.capital/.
[10]Official website: https://framework.ventures/.

- **Hashed**[11]

Hashed was founded in 2017 when four engineers from South Korea raised $700,000 on their own to make crypto investments. Five years later, after 150+ investments, their funds under management total over $4 billion and they do not have any external LPs.

Hashed has two big investments, one in decentralized stablecoins and the other in gaming and metaverse. Their investments include Axie Infinity, Decentraland, The Sandbox, Yuga Labs, etc.

- **Multicoin Capital**[12]

Multicoin Capital is probably the most rewarding venture capital fund. In 2017, the two founders of Multicoin Capital, Kyle Samani and Tushar Jain, launched this venture capital fund after the ICO bubble. Unlike the founding teams of other VCs, Kyle and Tushar had no experience working in investments, had not they worked in emerging startups, did not have establish any agreements. However, with their unique "outsider" perspective, they have "beaten" the VCs and gained a foothold in the highly competitive crypto space.

Multicoin's three main themes of investment are open finance, Web3, and non-sovereign currencies. Its typical investments include Flow, Fluence, LayerZero, Project Galaxy, etc.

- **Pantera Capital**[13]

Pantera Capital was founded in 2003 by Dan Morehead, former head of macro trading and CFO of Tiger Management. Pantera's global macro-strategic investments exceeded $1 billion and the Pantera Bitcoin Fund is up 57,500% since inception. Pantera Venture Funds has realized $125 million in profits from $23 million in invested capital in 26 companies. Its investments cover various aspects of blockchain

---

[11]Official website: https://www.hashed.com/.
[12]Official website: https://multicoin.capital/.
[13]Official website: https://panteracapital.com/firm/.

infrastructure projects, financial projects, enterprise projects, Web3 applications, etc. Its investments in the Web3 space are focused on the DeFi segment (including 0x, 1inch, Balancer, etc.), and it has not yet made a push in the NFT or chain tour space.

- **Paradigm**[14]

Founded in 2018, the young Paradigm has long been recognized as a Web3 head VC. Founded by Coinbase co-founder Fred Ehrsam and former Sequoia partner Matt Huang, Paradigm has received donations from three major universities — Yale, Harvard, and Stanford — along with the backing of well-known venture capital firm Sequoia. They believe that if the Internet has defined innovation in the past decades, the next decades will be defined by crypto.

Paradigm's portfolio includes well-known projects such as Uniswap, OpenSea, Starkware, Optimism, and Gitcoin, with investments ranging from $1 million to $100 million.

- **Polychain Capital**[15]

Founded in 2016, Polychain Capital is one of the first native crypto hedge funds to emerge, investing in numerous early-stage start-ups and protocols. Founded by Coinbase's first employee, Olaf Carlson Wee, the fund has attracted investments from a number of high-profile venture capital firms including Andreessen Horowitz, Union Square Ventures, and Sequoia Capital (Sequoia Capital).

The company has always believed that as the cryptocurrency ecosystem evolves, many different protocols will emerge to accommodate different use cases, hence the name "Polychain". As of February 2022, Polychain's assets stood at $5 billion, an increase of 125,000% since inception. Polychain Capital's portfolio includes Horizon Blockchain Games, Web3 Foundation, Coinbase, and others.

---

[14]Official website: https://www.paradigm.xyz/.
[15]Official website: https://polychain.capital/.

- **Sequoia Capital**[16]

Sequoia Capital was founded in 1972 and was the first institutional investor in old Internet companies such as Apple, Google, Cisco, Oracle, Yahoo, and Link. However, the early founding of Sequoia Capital has not been a stumbling block on the way to embrace innovation. The veteran VC is now also active in the emerging Web3 field.

In June 2022, Sequoia Capital launched two new funds with a total size of $2.85 billion in India and Southeast Asia respectively to expand its investment in the Web3 field. "Betting on the track is more important than betting on the racers", which is an incisive summary of Sequoia Capital's founder Don Valentine's investment style.

The projects Sequoia Capital bets on in the Web3 field include LayerZero, StarkWare, Polygon, Parallel Finance, etc.

- **Spartan Group**[17]

Spartan Group is headquartered in Asia, and its investment team has over 20 years of experience in investment research and capital management at top firms such as Goldman Sachs and Indus Capital.

In March 2022, Spartan Group announced its plans to launch a $200 million metaverse fund, which will focus on projects built in the metaverse, specifically supporting "digital ownership" of virtual worlds. Spartan Group's portfolio includes BitDAO, Arbitrum, LayerZero, Zapper and others.

- **Three Arrows Capital**[18]

Three Arrows Capital is an emerging markets-focused hedge fund manager founded by hedge fund managers Su Zhu and Kyle Davies. Founded in 2012 and headquartered in Singapore, Three Arrows Capital is focused on delivering superior risk-adjusted returns. Three Arrows Capital's portfolio focuses on the following sectors:

---

[16]Official website: https://www.sequoiacap.com/.
[17]Official website: http://spartangroup.io/index.html.
[18]Official website: https://www.threearrowscap.com/.

underlying public chains, DeFi, NFT, and gaming, funds, and more. In terms of volume, Three Arrows Capital's portfolio is second only to a16z. Its portfolio includes Axie Infinity, dYdX, Balancer, and others.

Note: Despite becoming a top crypto VC in just a few years of its existence, Three Arrows Capital has recently been affected by the unanchoring of the stablecoin UST and is facing a liquidation crisis with an estimated net debt of $1.6 billion. (Recently, according to media reports, a British Virgin Islands court has ordered the liquidation of Three Arrows Capital.

- **Tiger Global**[19]

Having kicked off its investment in the crypto space back in 2015, Tiger Global, a well-known hedge fund and venture capital firm, has been accelerating its presence in the crypto space since April 2021. According to Crunchbase, Tiger Global has invested in about 41 projects in Crypto so far, with 46 cumulative strikes across eight verticals and $7.17 billion in total project funding.

Tiger Global, a 20-something-year-old "tiger" that has captured tech giants such as Meta, Uber, ByteDaily, and Jingdong, is now killing it in the crypto space with its aggressive investment style and will be investing more in the Web3 space starting in 2022. Its portfolio includes Yuga Labs, Nansen, LayerZero, Polygon, and more.

---

[19]Official website: https://www.tigerglobal.com/.

Singapore University of Social Sciences - World Scientific Future Economy Series

(Continued from page ii)

Forthcoming Titles